Tony —

Hope you enjoy some of the customer care tips the old fashioned way —

Guy.

10/08

Return of the Body Snatchers

By Cary Blair & Ron Watt

© 2005 Airport Books LLC

ISBN: 0-9709632-2-X

Library of Congress Control Number: 2005927066

Cover Design and Illustrations by: Tom Glover

Airport Books LLC
17403 Edgewater Drive
Lakewood, OH 44107

Printed in the U.S.A.

Published in the United States by Airport Books LLC,
Cleveland, Ohio.

www.airportbooks.com

Acknowledgments

The authors would like to thank a number of people without whom this book would *not* be possible.

John Warfel had the common sense to have Cary Blair and Ron Watt meet, with a handful of others, at a "thought conference" that Cary hosted at his adopted home in verdant and oceanic Amelia Island, Florida. Much time was devoted to playing golf and socializing but even more hours were spent talking about business, marketing and sales issues. A bond developed, and the two talked about the possibilities of starting the book you now have in your hands.

John, the president of Westfield Financial Corporation, is the consummate sales and social person and he embodies much for which this book stands.

At the same time, gratitude is bestowed upon Paul Carleton, one of America's top investment bankers, who always thinks as a salesman and creative business person but also possesses people skills that are second to none. He utilizes these skills every day and "networks" many people who otherwise might not know one another. He is another key reason the authors met and became friends, and he, too, is a working example of the "right way" premise of their book.

Acknowledgment is extended to the gifted and diverse artists of *The New Yorker* and that wonderful magazine itself for allowing us to use their cartoons throughout *ROTBS*. And we want to give a tip of the hat to Tom Glover, the talented artist and friend who created the cover art and the original cartoons appearing within.

We'd like to thank Jeff Stitz, graduate student at the University of Akron, and Megan Bublik, of Watt Consulting, for their research work and other help in coordinating this production. And thanks, too, to Ron Watt Jr. for eagle eyes on the manuscript.

Authors and social etiquette experts Peggy Post and Peter Post; Fariborz Ghadar, director of the Center for Global Business Studies in the Smeal College of Business at Penn State University School of Business; and Susan Insley, executive vice president and principal of Cochran Group and former senior vice president of Honda of a America Mfg. Inc., were contributors to the book and continue to inspire the authors with their uncommon wisdom.

We also want to express our added gratitude to our wonderful wives who have been part and parcel over the years in all the successes and blessings that we have enjoyed, personal and otherwise. Karen Blair and Simona Watt, this one's for you!

Of course, we can't forget our original inspirations for this book. They include the 1956 movie *Invasion of the Body Snatchers* and its death-defying protagonist Dr. Miles Bennell.

We also want to thank the soulless, heartless, emotionless "walking dead" in that film and their progenies, the walking dead Poddies of today. Poddies lurk in cubicles in every medium to large corporation and in government agencies, non-profit organizations, schools of higher learning, religious institutions, trade unions, and trade and professional associations throughout America today.

Were it not for these gadget-headed and screen-addicted clones and clowns this book might never have been written.

C.B. & R.W.

Preface

Return of the Body Snatchers is a business book for students and young executives and their bosses, all of whom have high hopes for their careers in the "customer care" disciplines. This is for the generation of young people who started their lives with PacMan and have known nothing but high tech ever since. Inspired by the 1956 cult-sci-fi classic, *Invasion of the Body Snatchers*, our book examines in a spoofing way how the seductive and addictive characteristics of today's gadget world have possessed our younger generation. Their bodies and minds have been snatched up by the alluring qualities of today's hardware and software and the devices that make their programs so seemingly user friendly.

Our future business leaders have been caught and cloned and dropped into cubicles, behind screens to never see the face of a real live customer for ostensibly their entire careers. Instead of partnering with technology, they have bought in hook, line and sinker to Gadget World. Life in the Pod turns these clones into employees or public servants who are *weak, complacent, tedious, predictable, boring, loutish, anti-social, changeless and risk averse.*

We call them "Poddies" . . . People of Dullness and Drudgery Imbedded Everywhere.

This book describes the "escape route" from dullness and drudgery and internal focus back to time-proven, externally focused customer service—a lost art in the new world of e-mail, voicemail, cell phones, handheld computers and other abounding gizmos.

The book looks at ways to create a new partnership with technology for personal success, rather than being wowed by it all. Take it from the slightly older set, customers still like to see you, they like to be thanked personally for their business and they love to be treated in human and caring ways. Enjoy reading about "customer care" the old fashioned way, from two authors who spent their entire careers preaching customer focus to employees of their firms, to their clients and to the world!

"It's always 'Sit,' 'Stay,' 'Heel'—never
'Think,' 'Innovate,' 'Be yourself.'"

CHAPTER ONE

They're Back!!!

THEY'RE EVERYWHERE and you don't want to encounter them, because they don't want to encounter you—unless you're one of their clones, perish the thought.

They prefer to reproduce themselves and even you from the Pods in which they grow. The Poddies —People of Dullness and Drudgery Imbedded Everywhere—are ceaseless in their efforts to be weak, complacent, tedious, predictable, boring, loutish, anti-social, changeless and risk averse.

Slowly they sink into mediocrity and take you with them, and also the company. Change is not their game; thinking out of the box or the Pod offers no fulfillment. Their contentment lies in the Pod in which they reside. It's safe.

They are often bright, but they also are $150-an-hour people doing $25-an-hour work. They submerge

———

themselves into their devices that seemingly make them efficient but effectively do the exact opposite, lowering the value of the company. Salesmanship and communication are the last things they are interested in. That would scare them.

They specialize in avoidance, perfectly content to bury themselves into their e-mails, their voicemails, their BlackBerrys, Palms, i-Pods and ever-present cell phones. Mentally, they work in a closet instead of a ballroom. They work one day at a time. Visionary or futurist they are not.

Quite certainly, you'll discover them inside America's most successful companies and the worst ones as well. Curiously, you almost never find them in the arts—those people who paint, sculpt and photograph and those who play and write music. You can say the same thing about novelists and other fine writers. Journalists, though, are another matter. Talk about clones. Most are driven by the "who, what, why, when and how" school and many are too lazy or non-innovative to part from that box.

Clones also permeate sports. Think about it. Does every NFL football team—save one or two—look the same but in different uniforms? Their approach to the game is the same, predictable, boring. The one or two—most recently the New England Patriots— that part from the pack enjoy sensational success. Most NBA basketball teams look the same in their approach. All you have to do is watch the last two minutes of any game and that *is* the game—where one team breaks out and wins. The rest of the action often is just a bunch of guys running up and down the court, sneakers asqueaking. Exception: most any game that Larry Brown, Phil Jackson (nine NBA titles), Red Auerbach (nine NBA titles) or Pat Riley coached. Cerebral. Different.

Left unabated they'll surely take down even the best of companies—or make them ordinary at best. It happened at IBM, TRW and INA. Even the most creative companies such as Microsoft and Apple have their Pod people, people making a good deal of money but not offering a thought to the advancement of corporate energy.

In 1955 a movie starring Kevin McCarthy as Dr. Miles Bennell appeared on the big screen. "The Invasion of the Body Snatchers" is about a small town in California that encounters a curse where its residents one-by-one are taken over by extraterrestrials, emerging from giant seedpods. The residents' bodies are replicated in the pods and their real bodies die. The result is not much different than the walking dead. People without emotions, souls, hearts and love. Soon the entire town is overwhelmed with horror and it doesn't stop there. Dr. Bennell seeks to warn the rest of the world of the Pod people. The film to this day is a sci-fi cult classic. You should rent it. Its message will have some chilling underlying thoughts about what is actually occurring in America's companies, institutions, universities, legislative bodies, associations and religious organizations today.

Ironically, there are not enough Dr. Bennells in our corporations today who wish to warn top management, customers, vendors, shareholders, the financial community—and anyone else who needs to know—about the destruction the modern Poddies

---

are causing. Conformity is getting worse than ever. Check out the Big Four Accounting Firms. Big Law Firms. Big Management Consulting Companies. Big Banks. Big Government.

These facts are true in many areas of companies, not just in information technology. Poddies are brewing in HR, law departments, PR groups, operations teams, sales departments and where the original Poddies came to be—the financial departments. Lest we just not expose the current dilemma, there are many other examples of Podology that can be found in the past 100 years and even farther back.

One of the first Pod-creating concepts was the American labor union. Much good was done for workers in terms of safety, pay, benefits, working hours, and child labor legislation as a result of the union effort.

One of the first unions was the Knights of Labor, which was founded in 1869 in Philadelphia for the garment workers. Later came the Federation of Trades and Labor Unions, which was re-designated the American Federation of Labor in 1886, the AFL-CIO that we know today. Throughout a multitude of divergent industries—from steel production, coal and ore mining, and transportation to meatpacking, manufacturing, construction, the arts, and sports—unions found their way.

Now the bad side: Labor unions have created a philosophy for workers where productivity and efficiency are not necessarily important. Wages are based en masse, as opposed to upon an individual's skills, work ethic, achievement, efficiency and production. Union workers typically are not allowed to negotiate their own contracts for the work they perform. As a result, motivation goes lacking, creating a paralyzing-like environment of *People of Dullness and Drudgery Imbedded Everywhere (Poddies!!)*

Further, the Clayton Act of 1914 exempts unions from anti-monopoly laws, allowing union officials to force out independent or alternative employee bargaining groups. This power allows unions to step away from the foundations of capitalism on which the United States was built and, instead, work from a more socialistic model. If it looks like a Pod, and sounds like a Pod, by gar, it must be a Pod!! Let's all think and act the same.

Here's some more food for thought. Unions have become more powerful than the government when it comes to construction projects. They have dictatorial rights beyond that which most of us enjoy. Based on the project labor agreements, at any level of government, the entities of government are required to award contracts to unionized construction employers for major projects such as highways, bridges, airports and, of course, government buildings.

Not only can unions work against the corporation and government but the union members themselves as well. Through the landmark National Labor Relations Act (NLRA) of 1935, unions have exclusive negotiating rights with employers. In other words, the head Pod dictates what all the other Poddies are going to have for dinner, and they had better like it. And the unions have the power to collect union dues from each of their Poddies, whether the individual Poddie wants to give or not.

Most labor union contracts ensure that the constituents have Election Day as a paid holiday, so they can make it easy for their Pod people to vote for the candidates the unions' approve.

So you can see, Poddieville just didn't start a minute ago. It's been with us a long time in one form or another, and as for unions, since the 1860s.

But let's not overly lay it on the labor unions. Seed Pods have popped up all around us over the years in every nook and cranny of American business, government, institutions and organizations, and they are getting ever more fertile as days and months go by—everywhere!

They take the lifeblood, the soul, the heart and emotions out of us, and we had better stop them before they stop us.

This should give some of them Poddies a lift!

CHAPTER TWO

Who are these 21st century body snatchers?

THE ORIGINS OF THE PODDIES are interesting to behold. Where did they come from? Who are the modern day Body Snatchers that are producing them?

As good as they are and as freewheeling they profess to be, certainly the most obvious Body Snatchers are the high tech hardware and software firms. Characterized as the most innovative companies in the world, they ultimately produce products and services that have little "wiggle room" in their use. Everyone must follow the script, everyone must act the same. Either follow the documentation or end up in the never, never land known as the "help" desk or have your password stamped "invalid," electronically of course. Manuals and documentation make the user controlled and compliant.

Certainly you can count on Microsoft and Apple to lead the way in cloning their users. Think about it,

Word, PowerPoint, Excel, Outlook, and other Microsoft programs are just one culprit that puts everyone—or most everyone—in the same box. How do you think outside of this trap if you are using their software? You don't, rules are rules!

Look across the office to the so-called creative departments and they are booting up their Macintosh software. They are in the same boat—but a different box. More Body Snatchers are making you a cloned user of this highly possessive software.

If that were not enough, Microsoft and Apple software do not communicate with each other, lest one or the other lose their identity and their power over "their Poddies." So never the lines shall cross. Separate troops caught and cloned. Pretty nifty, huh? Hundreds of millions of people in these massive armies confined to classrooms, cubicles and homes across the world. All of them fixated on one or the other system, coached and coaxed by one or the other system.

Not to be lost when identifying the "big boys in body snatching," we all know the names SAP, PeopleSoft and Oracle—the world's biggest proprietary business processing systems. These systems are now used almost exclusively in manufacturing and services industries, from operations monitoring to payroll. The installation and documentation of these massive systems created a home for thousands of consultants even before Congress enacted the "Consultants Security Bill," known as The Sarbanes-Oxley Act of 2002. With that boost they now run rampant as never before.

It would be rare for any company to install either SAP, PeopleSoft or Oracle software systems without an army of consultants to ensure that programs are installed the same way as every similar installation, good or bad. These Body Snatchers have seen to it that competitive flexibility is a no, no. If you try to get a little innovative, your very existence as an entity could be threatened. Just ask the people at Hewlett-Packard about their installation experience with SAP. They, too, have joined the cloned flock of users who now make no changes and are cloned to look like all other users, their competitors.

George Orwell in his epic book, *1984*, displayed an imagination about Big Brother taking over our lives. But the reality could prove to be worse than Orwell conjured, and it came just a few years later!

How about Lotus, which means in Greek "a fruit that is supposed to induce a dreamy languor and forgetfulness." Here you have another star in the current line of Body Snatchers. They have literally created a new means of electronic communication—Lotus Notes—that has just about replaced formal business letter writing. Now we have endless works of e-mail, strangely written, poorly thought through, curtly delivered and often misspelled. And this goes on and on until the messages finally lose gas. It's working! Everyone has forgotten how to communicate.

"You've got mail."

Peggy Post and Peter Post in their book The Etiquette Advantage in Business (Harper Collins) aptly observe that today most business executives bemoan good business writing as a dying art. In truth, good writing is not quite ready for burial but is definitely in need of resuscitation. The new reliance on lightning-fast electronic communication—e-mail—has given rise to a breezy style that has little to do with the formal composition of traditional business letters and reports.

E-mail allows us to avoid. Avoidance of human contact and building face-to-face relationships are the external by- products of e-mail. Think about it. Hours and hours each day devoted to this drivel. You never have to pick up the office phone or call outside the office when you are under the control of the e-mail devil in your computer. Why bother? You are under the system's control and, of course, the system makes you compliant and poorly productive, sucking your energy as an individual human being with his or her own persona.

There is no question that the authors of Lotus Notes must be identified as modern day "Body Snatchers" who have created Poddies who have a dreamy languor and forgetfulness by the millions.

Communications . . . noun, a giving or exchanging of information, signals or message as by talk, gestures, or writing.

Today's "communications" companies have effectively destroyed the typical meaning of the word. Traditional telephone companies lead the parade in poor communication techniques; the very companies that should be the teachers of good communications stick us into automated answering systems and automated service centers where electronic voices try to talk to you. Have you ever slipped and actually tried to respond to one of the better electronic voices?

SBC/AT&T, Verizon/MCI, Sprint/Nextel and Qwest have snatched us up and are trying to tell us how to communicate our service desires by punching one, two or three on our telephone pad, never to hear a real human voice down their hard wires again. They have cloned us into believing that this kind of service is exactly what is good for us, whether we like it or not. What in the world do the elderly do when they encounter a problem with their telephone or need to talk to the company about a service problem? They must ask their grandson or granddaughter to take care of it for them, kids who grew up in this mirage of technology.

Cell phones of course are the rage and soon you will be able to use them on airlines at all times, not just before takeoff and after landing. Just think how much you enjoy your seatmate's cell conversations today, before and after the flight. You'll have a ball listening to them for the entire flight! This will be an all-new way to "hijack" the plane and you won't be able to anything about it, gripped inside the tube as it flies through the air. You'll be on the Babble Jet!

Bill Peltola, an executive with the aviation-communications firm AirCell, wasn't off base when he warned of noise problems if cell phones become commonplace on airplanes. "It's not a trivial issue," he told USA TODAY last month, adding in jest: "Maybe we'll have little plastic cones drop down from the ceiling when you talk on the phone."

In-flight cell phones? Why do it at all?

There is already so much "air rage" occurring everyday. Air travelers are anxious, cramped, tired and hungry. Why add salt to their multiple wounds?

Why merge two of the most combustible realities of daily life—public place cell yell and the frustrations of air travel? Airplane cabins are already rife with potential confrontations between generally courteous folks who are pushed to the brink by hostile conditions. Add in-flight cell phone conversations? No way.

The outcry against cell phone noise is widespread. When the public research group Public Agenda asked people their thoughts on a law to ban cell phone use in public places such as museums, theaters and restaurants, six in 10 said it would improve people's behavior.

According to another survey by Public Agenda and Travelocity, 65 percent of airline passengers believe rudeness while traveling is a serious problem. Any peace and quiet on airplanes would be shattered by in-flight cell phone calls. Really, what is so important that it can't wait until the plane has landed? The only reasonable exceptions are extreme emergencies.

I'm the first to admit to being a multi-tasker. I'm also a frequent business traveler who views the relative peace and quiet in flight as one of the positive features of being on an airplane. Honest. Even if I find myself in a middle seat in coach class, I'm able to read, write and think. It's my "quiet time"—an ever decreasing and precious commodity in this day and age.

I'm all for progress and innovation. But how would widespread, noisy in-flight cell phone use make our lives better?

I just don't get it. Why create a new way for people to be rude to each other? We, the public travelers of the USA, just don't need this innovation.

—from Peggy Post in USA Today, August 4, 2004. Ms. Post is an author and executive director of the Emily Post Institute, which provides individual and business etiquette advice.

Although the cell networks add an incredible means to communicate, most of the time they do lock us into their domain, their way of doing things, or else. This sounds like one by one we are being taken over by extra-terrestrials," just like the movie, but this isn't science fiction!

It seems that being an individual, with his or her own resources, becomes harder and harder. Being able to break out from the mold of the machines becomes more difficult. We now add BlackBerrys, Handspring Treos, Axims, HP iPaqs and every other PDA to the mix. We desperately rely on Bluetooth and Wi-Fi wireless protocols.

And, of course, your voicemail can answer my voicemail and on and on into oblivion. No need for an executive assistant or secretary to serve as an intermediary either. The device manufactures chortle at the possibilities. For us, we lose a piece of our human identities, personalities and unique qualities. We all are becoming one, and not in the best ways possible. My god . . . *"The Return of the Body Snatchers!"*

Although certainly not in the tech world, the last major group of Body Snatchers and promoters of "all ducks in a row" have to be our lawmakers, regulators and judges who have been elected or appointed in the last two decades.

Our congressmen, both state and federal have seen to it that every abuse of the "system," small or large, be overcorrected with new legislation over the last 35 years, including the Occupational Safety & Health Act of 1971 (OSHA), the Employee Retirement Income Security Act of 1974 (ERISA), the American With Disabilities Act of 1990 (ADA), the Health Insurance Portability and Accountability Act of 1996 (HIPPA) and the Sarbanes-Oxley Act (S-OX) of 2002.

Compliance is the name of the game today in business and it is rare for a CEO to consider any major operational decision without first looking at the consequences of these major pieces of legislation affecting business life and the concomitant expense of compliance activities. All of the fun of being a CEO and running a company is gone, legislated away by the Body Snatchers, taking away most of our competitiveness and making everyone of us look alike. No wonder we can't compete around the world! Ask the European business community; they think we have gone absolutely mad.

Regulators have become the kingpins of compliance thanks to the plethora of new law and regulations dealing with business miss-steps by a few senior officials of a relative handful of companies. A few companies screw up and the rest have to pay for it. This has been common for the past 100 years in business in the United States. In states like California, New York and Florida, regulators continue to be quite harsh in bringing business "in line" while they personally use the media bully pulpit to advance their careers in public service.

The third of the terrible trio of Body Snatchers is the increasing number of judges who are not satisfied with interpreting law but insist on creating new law from the bench, a no, no in the separation of powers in the federal and states' constitutions. Rather than expressing opinions based on the letter of the law, many judges have found it simpler to make very liberal interpretations in cases to "protect the public," resulting in actions never anticipated in the original draft of the law at the legislative level. The result is a business atmosphere that is quite unfriendly, more compliant and certainly less competitive.

In some cases, we can elect to ignore or try to change the overtures of Body Snatchers if they are human. Members of Congress and many judges can be voted out.

Regulators are tougher to dethrone because of the size of their Poddie staffs and their mandates for compliance. The power of bureaucrats is endless and growing stronger by the day. Politicians come and go, but the bureaucrats last and last and become stronger, with an ever-increasing staff of Poddies: *People of dullness and drudgery embedded everywhere.*

But, still, there is no end to the abuses and plain idiocy of the legislative branch, particularly at the state level. In Ohio recently a legislator introduced a bill to curb professors from speaking out on any issues other than the subject that they are teaching.

Among other devices designed to control universities, Ohio State Senate Bill 24 states "faculty and instructors shall not infringe the academic freedom and quality of education of their students by persistently introducing controversial matter into the classroom or coursework that has no relation to their subject of study and that serves no legitimate pedagogical purpose." So, basically, if a professor teaching Shakespeare speaks about a current event, he would be violating the law.

—The Buchtelite, editorial from the student newspaper of the University of Akron

This bill is designed to withhold funding for state universities that allow professors to expand and elucidate beyond their basic curriculum. That is, "don't be a human being, a person of feelings and emotions. Don't have opinions, don't think, don't interpret. Stay with your boring topic and don't cross the fence." Bizarro!

The very meaning of university and college is togetherness and exchange of ideas. In Latin, the words mean "whole" and "partnership." What kind of partnership are certain pillocks in the Ohio State Legislature trying to invoke here?

But beyond goofball judges and legislators, stifling Acts of compliance and regulation and the bureaucrats that carry them out, self-perpetuating big law firms and accounting firms, and corporate Poddivilles that are heaped upon us, the greater challenge is technology run berserk.

New technology is the Trojan Horse. Gadgets and screens, keyboards and voice response, wireless remote and picture phones, disks and digital, e-mail and voicemail, hand-held or laptop, Internet and Extranet, all in such a friendly package. This Trojan Horse is such a nice gift to us and the world.

Maybe this *is* science fiction, just like the movie, *Invasion of the Body Snatchers*. It could be an out-of-body experience. The mechanical soldiers inside the Trojan Horse have quietly snatched our bodies with hardware and software that we admire and without which we cannot function, like a drug. Its power is addictive and possessive but relatively inexpensive. We can do things with our gadgets *now* that we never dreamed of five years ago. The world is at our

"I'm sure it's all right. It's a <u>horse</u> you have to worry about."

fingertips! We can't get enough of it. Gadgets abound to make our day.

Welcome to your cubicle where people of dullness and drudgery are imbedded everywhere. We call them Poddies!

*   *   *

Many of our automobiles can be pinpointed wherever they are by satellite systems. We are not far from the same with cell phones—always Big Brother can find us wherever we are. Privacy? Always easy to find or reach unless you go to a remote area of a remote State or to a third world country. More and more the devices find us and control us, rather than the reverse.

There is no doubt that the cell phone, computers and PDAs have added an incredible means to communicate, but at the same time they lock us into their domain, their way of doing things, or else. Being an individual, with his or her own resources, becomes harder and harder. Being able to break out from the mold of the machines becomes more difficult. Being able to socialize humanly, in person, takes less and less of our time. Not to worry, your voicemail can answer my voicemail and on and on into oblivion. No need for an executive assistant or secretary to add a human element as an intermediary either. We have all kinds of voicemail devices, at home, at the office, on our cell phones, and PDAs, such as BlackBerrys, Palm One Handspring Treo 600s , MXP 220s and BenQs. We are device drones!

E-Mail and Voicemail can be indispensable when you wish to do nothing but just respond.

CHAPTER THREE

They're everywhere you look!
All the Ducks—errh—Pods in a row

THE ORIGINAL INTENT of hardware and software companies was to create very efficient, leading edge solutions to current business problems. Their mission statement probably did not mention the capturing and cloning of users, utilizing possessive, alluring technical advancements and gadgets. Nothing was ever said about the ultimate result of creating *People of Dullness and Drudgery imbedded Everywhere* (Poddies) in our corporations, institutions and legislative bodies. But they have.

Can you imagine that our giant communications companies actually think they have gotten it right? These super communicators believe that we are all electronic wizards who enjoy electronic service centers and the fun you can have pushing key after key on your telephone just to change your billing address. They are trying very hard to make us just like the Poddies that are imbedded everywhere in their companies.

Being institutionalized must be good for us, because never in our history have we allowed ourselves to be forced to act in such confining ways by our Big Brothers, whether it is government, the private sector or institutions like universities or churches.

Entrepreneurship is just about dead, a more and more a figment of the past!

The cost to be different is just about unaffordable. Try to get a bank loan today for a new venture, a new idea, and a new business. Lenders know that you are destined to fail 90 percent of the time just because of new government compliance costs.

Just like the science fiction tale, our bodies are being snatched, caught and cloned and we end up as weak, complacent, tedious, predictable, boring, loutish, anti-social, changeless and risk averse people.

Poddies are being created everywhere, all marching to the same tune. All the Ducks—errh—Pods in a Row!

Abuses at the top of American businesses have occurred cyclically since the history of private enterprise. The latest round involved Enron, WorldCom, Adelphia and a few others. Less than one half of one percent of the private and public corporations have created a real mess for the rest of us. True to the past, following each corporate miss-step, overbearing legislation such as OSHA, ERISA, ADA, HIPAA, and S-OX follows. Heavy regulation creates fields of Poddies everywhere you look and the latest is no exception.

The Sarbanes-Oxley Act of 2002 (S-OX) was enacted by Congress to deal with the latest round of corporate abuse and has brought the 99.9 percent of good American businesses to their knees in terms of cost of compliance and levels upon levels of financial checkpoints.

Executives of large publicly held and privately owned companies and smaller organizations now *really* have the Big Four accounting firms and their spin-off consulting firms to deal with. Not much choice.

When they were the Big Eight at least there was some differentiation among the accounting people, simply because of competitive forces. Then they went to the Big Six, and Big Five and now the Big Four. Maybe eventually to become the Big One! No matter, thanks in large measure to Senator Sarbanes and Congressman Oxley, they have become conditioned to do things exactly the same way, with the same kinds of people and the same mindset.

Same daily routines, cloned trainees, same regulations to enforce, same pricing to the client, same dress code, same company cars, same Palms and BlackBerrys, same software running on their computers.

Clone after clone. Partners and associates could switch from one firm to another and it wouldn't matter. They would require you to do everything the same way in the same personality but under a different marquee. These firms have to be used for certification and sign-offs. They like that! You will hear it referred to quietly as being "Enronized," as they become smugger, more costly and more dictatorial!

In the Midwest we see countless publicly owned middle sized banks, investment companies, manufacturing organizations and others that are now feeling the cost impact of compliance with SO-X and its onerous Section 404 regulations. This financial controls section must be tested and certified by "The Big Ones" every year. One big slip-up in the hundreds and thousands of internal controls and "The Big One" denotes that you have a "material weakness" that must be conveyed to shareholders in the annual proxy statement (10K); then goodbye stock price!

It would not be unusual for these stellar institutions, who have never been suspect to regulators, to be incurring millions of dollars in implementation costs of this new regulatory layer. Now, they must divert financial and auditing staff from important other duties to supervise a herd of consultants and additional employees.

Small and middle-market public companies are taking it on the chin with over-zealous regulation. These continuing costs rob the company of valuable profits for expansion, new product development, reinvestment in new operations, and generally result in a reduction in shareholder value over time.

What we are seeing is the never changing response to abuse in the marketplace by regulators, more laws to comply with and more departments and another layer of bureaucracy.

It is always the same: no thinking outside the box. What would have been wrong with clamping down on current government bureaus, departments and regulators to simply do their job because most of the laws and regulations to handle these abuses already existed??!!

—*Cary Blair*

Right down the street from the Big Four accounting firms are the just plain Real Big Law Firms. Things have been rocky for these guys and girls. Some of them have actually gone out of business and some have had to merge to stay viable. Little did they know that S-OX was coming to their rescue. The strong shall inherit the earth and that is exactly what the remaining Real

"The men are excited about getting to shoot a lawyer."

Big Law Firms saw as their Eureka. Racing down the track and through the tunnel came S-OX.

Sarbanes-Oxley pumped the needed lifeblood into these blokes and now they have joined the Big Four in interpreting, analyzing, strategizing and confounding clients across the country. Another bottomless trough of dough has been found! What a godsend! And these are defense lawyers, not the plaintiff bar so resoundingly known for finding fertile causes year after year.

S-OX and the current spate of laws, regulations and Acts now being leveled at business should keep the Real Big Law Firms going for centuries. And the same time S-OX will push many of us into a slow and agonizing spiral of more controls and compliance, snatching the lives out of our corporate bodies as they are being captured, cloned and controlled.

Body Snatchers are everywhere you want to inspect. Everywhere.

Our new round of future business leaders are graduating from today's prestigious B-Schools all lined up like *ducks—errh—Pods in a row.* Things go by the book here, or else. National certification is paramount for enrollment numbers, so the syllabus looks strangely alike from school to school. Mainly led by professors who have no practical business experience, their graduates reflect a by-the-book academe attitude consistent one to the other. The profs might take on an assignment here and a research study there, but how many times have they had to run their own business, a business of some sizeable merit?

When you are in the tomb of academe, why would you want to fly around in the land of risk, chance and potential failure? The profs would rather the students try and fail on the job than in the classroom. Why experiment in the classroom—*not* doing it "by the book?" Why suggest that doing it by the book may not be practical or cost efficient? Why create? Why think? Why make extra work?

The students can "experiment" when they reach their corporate enclaves. But for now, by the book is the way it shall be, and await the new class next year.

Without meaningful internships, what do these kids know? But most are entering the workforce at astronomical salaries and benefits. Would they know entrepreneurship if it barreled them over the knees? Are they at all valuable to companies in their early careers? Can they bring thinking "outside the box" to the table? The corporation gets very little "lift" in return until some begin to get the picture during the ensuing years as the *real* training commences.

The skill of face-to-face communication and relationships has been lost in the B-School curriculum. The deans and professors must think that electronic voices and gadgets get the job done. There is no question that the Body Snatchers have created willing Poddies in the world of academia. Customer care skills went by the wayside long ago, undoubtedly because these skills cannot simply be taught out of a book.

Don't look for sales courses in the B-Schools. Don't look for courses in communication, advertising or public relations. Less than a dozen out of hundreds of business schools attempt to teach sales and other communication skills to their students. These "customer care" courses ask students to develop interpersonal adequacies, something that causes anxiousness among Poddies in the professor ranks.

The University of Akron is one of a few universities to offer collegiate sales education. Through the Fisher Institute, the University of Akron offers at the undergraduate level a major in sales management, and a certificate in global selling. The graduate sales program offers a concentration in global sales management. The university claims there is no other business program that offers such an extensive array of sales education programs. In fact, out of 2,500 universities in the country, the University of Akron is perhaps only one of five to offer a baccalaureate degree in sales. Worldwide, there are fewer than 10 universities that offer a major in sales to their undergraduate students.

The mission of the Fisher Institute for Professional Selling is to enhance the image of the sales profession— to promote professional selling and sales management as a rewarding lifelong career discipline, to provide high quality instruction to both students and established executives, and to conduct research that advances the field of sales.

*   *   *

Talk about the need for salesmanship, here's an institution that could use some help.

"Now, this complete, all–in–one model has a thirty–nine–tube television receiver, equipped for both black–and–white and color reception; AM and FM radio; a record–player geared for 3 1/3 r.p.m., 45 r.p.m., and 78 r.p.m.; automatic record–changer; the latest thing in a wire recorder; and this large, roomy cabinet at the bottom, in case anything new is invented."

Across the U. S. the Catholic Church is losing its younger members, many of whom were educated in elementary schools, high schools and colleges run by Catholic orders. Still, the younger ones are falling off, as statistics indicate. This phenomenon is a blow to the future of the parishes and the schools, many of which have been forced to close or consolidate. Most of the problem has been caused by the rote nature of religious teaching and, more so, by the clone-like approaches to handling something that could be simple, special and graceful—the Sunday mass. Also at issue is the intransigence of the Church's cloying hierarchy, seemingly unable to understand the "needs" of its customers.

But the Catholic Church does not have clearer instincts for body snatching than other religions— Christians and non-Christian alike. The overbearing, controlling nature of our religious institutions has caused a cascading drain of young people who are just not buying dogma.

Many of these young adults, many of them trained in religious schools, are finding ways to fulfill their religious needs through other means of worship. They may choose a walk through a beautiful metropolitan park where nature and God's work abound. Or they may try simple meditation, which does not require the recitation of rote prayers.

Whatever, they are going about it "their way," while the institutions are still bent on trying to make them Poddies. But more and more, young worshipers simply will not put up with the Podheaded nature of religions and simply don't want to become clones in the name of God.

*   *   *

Take a good look at the chambers of commerce and other business trade associations that once were fanatical about their causes and struck fear into politicians and media who would cross their path. Those were the old days when they were trying to make a name.

Today, their game is bureaucracy. They comprise people protecting their jobs first and foremost, and the result is that nothing truly is getting done. Employees are keeping their jobs and blithely making themselves believe, their boards of directors believe and their members believe that progress is being made. But mostly it isn't; risk taking has become passé—making waves might get you drowned! Today, these chambers and associations, dating back to the 1850s, are perfect for the Poddies—accountability for business gained, economic development defined and tangible lobbying efforts are often foggy at best.

The State of Ohio, which was key to the re-election of a Republican President in 2004, is a perfect example of a business friendly state gone bad, right under the eyes of the chambers of commerce and the manufacturers and transportation associations. Once the most business-friendly state in the nation under former Governor Jim Rhodes in the 1960s and 70s, it is described as the "worst" today among major industrial states. Why haven't the Ohio Chamber and the big city chambers taken a clear and active role in these problems? They haven't because they are staffed these days with Poddies—*People of Dullness and Drudgery Imbedded Everywhere.* Consistent with having their bodies snatched, they have turned these once evangelical organizations into something mirroring *weak, complacent, tedious, predictable, boring, loutish, anti-social, changeless and risk-averse. All the Ducks—eerh—Pods in a Row.*

In Ohio, companies, jobs and people are running out of the state in record numbers, while taxes and regulations continue to wind upward. Wouldn't it seem logical that the Ohio Chamber of Commerce would take an active role in reversing these problems? If it is, we don't seem to know much about what the

Chamber is doing. And whatever it *is* doing simply isn't working.

In the 16 years old Jim Rhodes was in office, when he was the Governor in the 1960s and 1970s, the exact opposite was true. The state was on a roll with countless trade missions to Europe and the Far East, attracting Honda to build its plants in Marysville, for example, and fostering much exporting from Ohio companies to countries around the world. The Chamber and the Governor were no people to mess with. They were evangelical!

As with chambers of commerce, trade and professional associations are natural breeding grounds for Poddies. Aside from lobbying our cities, states and the federal government, what exactly do they do?

An association may present 400 technical papers at its annual convention, but what does this all mean, other than sating the egos of those presenting? One has to look long and far to find lasting, meaningful results. The staffs of these institutions, though, just keep hammering away with their nonsensical work, and largely nothing happens. But they are safe and happy and unchallenged to produce anything of true merit for their respective industries and professions.

An executive director might make a few hundred thousand dollars a year, and his or her staff sixty thousand to a hundred thousands dollars a year. They supposedly produce services and benefits for their large member bases, but one has to only wonder what these are and exactly how they help the constituents and industries they represent. Our guess is the real deal is the husbandry of people who can't get jobs gainfully in real business.

Podology swarms around us. And just like the old movie *The Invasion of the Body Snatchers*, they're coming at us in profligate numbers and dimensions in the darkness of night and the murky bowels of the Pod by day. This makes George Orwell's *1984* Big Brother figure look mild by comparison, doesn't it?

But, unlike the movie and George Orwell's *1984*, there is a way of escape and we shall share that with you later in this book. Have faith, you can win and be the distinctive person you can be.

Men [and women] become susceptible to ideas not by discussion or argument, but seeing them personified and by loving the person who so embodies them.

—Lewis Mumford, 1895, The Conduct of Life, from Leonard Roy Frank's Quotationary (Random House Webster's)

"No, the computers are up. <u>We're</u> down."

CHAPTER FOUR

Whither the executive these days?

THE PODOLOGY, UNFORTUNATELY, can reach upward into the top levels of management. The current younger generation of executives is the first to grow up with computers and computer games from birth on. These are the same men and women who devoured PacMan as children and went on from there. They know no other method. But does this make them significant contributors to their companies? Maybe not. Maybe they are destined to making their companies ordinary . . . weak, complacent, tedious, predictable, boring, loutish, anti-social, changeless and risk averse.

It is not atypical for everyone gathered at meetings today to have a computer or other gadgets in front of him or her, even the CEO. We know this saves a lot of paper, makes one look so efficient. But from observation, it seems that these people spend more time in love with little machines than they enjoy conversing with their compadres. Then, they leave the meeting and go right back to their screens. We're afraid that when these guys get to be in an older generation they will be confronted with strange eye diseases and hand problems not known so much now but surely coming to the forefront insidiously later. And when this comes to bear, they will all but have lost any social skills they ever possessed. Not much fun to be had there in assisted living when you might not be able to run your gadgets and actually have to talk with people.

But now as computer whizzes and gadget geeks they imbed themselves in their little instruments. Maybe they believe being more isolated breeds thoughtfulness, creative solutions and awareness of information for problem solving.

How many PowerPoint presentations have you ever seen go well? Thank God for administrative assistants!

Here the machine takes over the meeting. Some sort of *Star Wars* production tries to occur and often the presentation looks more like a falling meteorite. Something goes atwitter and the prime message and sub-messages go into the ether, only to take the participants' time but to not leave their mark. The tools take over and the consistency of the day goes out the door. Maybe there is an easy way to make these points – and maybe it finds its way on a few pieces of paper. Thank God for handouts, which often are more effective than some weird motion picture that makes all the headlines, streamers, screen-in-a-screen, cutaways and all the other crazy stuff you see dashing and darting on CNN, at one time, seem quite normal. Where is Walter Cronkite when you need him?

Here's another problem:

So many management people have taken over their assistant's job, if they even have an assistant. The manager writes volumes of e-mail each day, takes care of voicemail on his or her land line, on the cell phone, on the home phone. Makes his or her own plane, rental car and hotel reservations and gets driving instructions from Google or Yahoo. All this in the name of efficiency and self-sufficiency. But all this takes time. And on top of that it is $25-an-hour work for people who are paid $100, $200, $300 an hour or more. Efficient???? Leadership????

Poddies emerge when they are least expected to do so. Some actually make it out of the middle management Pods in which they reside and reach upper management to further contaminate the company with Pod-grown isolationist ideas. They are going to make their companies *weak, complacent, tedious, predictable, boring, loutish, anti-social, changeless and risk averse.*

But we as a society asked for all this. Parents taught their kids no social skills. Their kids grew up in front of screens—TV or computer. These devices and related gadgets, such as BlackBerrys, Palms, i-Pods and MP3s, are endemic to their ways of life. But socialization may not be. Nor may be much creativity.

Home schooling may have added to this dilemma as well. When you have siblings and maybe Johnny down the street in the home school, there is not much chance for socialization. We think that some home schooling parents may actually be selfish. They just take vacation whenever they feel like it, not having to adhere to regular school schedules.

And can their kids chat up someone they have never met before? Are they going to be well-formed adults? Are they heading for the Pod? You bet, just another new herd of closet executives.

Even Microsoft and Apple, two of the most successful and creative companies of all time have produced a generation of Pod-Heads in their own organizations. In these cases, the Poddies and upper management like things pretty much the way they are. It's similar to a Poddie army, with a few generals, majors and colonels running the show but with the real insidious Poddie soldiers sitting rampant just below the surface. The Poddies are happy with their McMansions and two SUVs and the entire rote of their lives. Throw them a new idea or concept and they start shaking like a dog abandoned at roadside. Software charts their course and makes them clones.

And now while kids are growing up in these families they are becoming even more Podified with all the gadgets they have at their beck and call. From early childhood, through acneville, on into college they are taught to have most every thought and communication through these devices.

But we have to give credit where credit is due.

A visionary concept for the colleges has been their growing creation of "cyber cafes," where communication, problem-solving and making new friends becomes a way of life—it's high tech socializing. We're not talking about your typical chat rooms here. Sometimes you have to go with the flow, take advantage of technology and socialize at the same time. Still, even this is difficult for natural-born Poddies.

This conditioning and brainwashing carries them off to college where they can sit in their dorm rooms for hour upon hour with wired or wireless depots hooking them up everywhere around the world through the Internet. *Everywhere* other than socialization with their peers in ways in which humans were designed to function.

The result is a kind of walking dead, where the Poddier you are the more your fate is sealed. They eventually meet other Poddies, become partners, get married and have Poddie kids whose only social outcome is playing soccer. The cycle continues in a slow but vicious manner. We are breeding anti-social closet clones who will eventually work in Pods and settle into a life of drudgery and non-accomplishment, frightened of new ideas, concepts and directions. They don't serve their organizations well because they are bred to not think out of the box. It is always safer to stay inside of it, they think.

I was on the airplane the other day and said hello to my seatmate on the other side. After the pleasantries, while I started reading the Journal, I noticed that this fellow, within minutes or maybe seconds, fired up his computer. He had told me he was vice president of sales for a major consumer company whose name you see every day in your household. After a while I asked him how much time he spent on the computer vs. spending time with his key retail customers or his sales force? At first, he said he probably spends two hours a day on the machine and six or seven with people. Unfortunately, pausing for a moment, he admitted that the reverse is probably true. Things are not as efficient as he might have thought they were. And sales and customer relationships are not what they could be, or should be.

—Cary Blair

"Human Resources."

CHAPTER FIVE

Attack of the HR people.

HAVE YOU MET A HUMAN RESOURCES PERSON who was truly happy? All so grim, discontented and usually angry—and all this gets directed at you. They look at government regulations so much that they actually begin to think these are good for their organization.

Several key Federal Acts empowered these rascals. One was the American Disabilities Act of 1990. Another was E.R.I.S.A, from 1974. We don't quarrel with the many merits of these Acts, just some ridiculous side effects, such as in the case of ADA, requiring companies to provide special chairs for obese workers. Maybe a good training room and some nutritional guidance would better serve the employees and the company.

Recently, the Federal Government suggested that ATMs all have Braille keys—even the ones you drive up to. There are not too many sight-impaired people who enjoy driving that much in the first place, so is this the overkill or what? And what of the extra cost of re-engineering the drive-throughs with Braille pads? Whether these will be retrofitted for the sight-impaired and perhaps for the hard-of-hearing is still up in the air. Or will these regs only apply to new machines being installed? Stay tuned.

Often it seems that common sense is decidedly in violation of the regulations brought forth through this type of legislation.

One of the great growth areas of companies and institutions today is the HR department, some time ago simply called the Personnel Department when it was more user friendly and a lot smaller. That used to be the most friendly department of all, always championing the employee. Now their descendants in HR have almost put a new meaning to the word clone as they massively emerge within their Pod and think *compliance, compliance, compliance!!*

They are typically instructional, dictatorial, self-absorbed and devoid of the human condition, the irony being that the word "human" is in their job description.

Their M.O. is to categorize and characterize everyone by standards. Those would be: standard job descriptions, standard salary surveys, standard salary guidelines, standard performance assessment tools, standard entrance interviews, and lastly, and perhaps fortunately for the employee, standard exodus interviews, the only good of which may be the word "exodus."

Everyone is in the same boat, good or bad. Every human is the same. No special characteristics, talents or abilities here. No Technicolor. It's all in black and white.

The HR people think everyone must be like *they are*—vanilla or chocolate. No Rocky Road. No Very Cherry Jerry.

With increased federal and state government regulation of corporations, these Human Resource Pods have come into a life of their own. In some companies just about nothing is done administratively without getting the HR people involved. And these Poddies have specialties. They come at you from their Pods in packs and from all directions. Of course, the vast work they conduct is through e-mail and voicemail and through long policy statements delivered by FedEx that make lawyers work look like child's play.

HR people get conditioned by the regs. The HR Poddies often do not know what the words "sales" or "a reasonable profit" mean. They are cost and compliance factoids that simply get more expensive as they spread. Even top management has trouble controlling them because, with Sarbanes-Oxley now heaped upon management, the seeming need for HR Poddies is greater than ever before. *Weak, complacent, tedious, predictable, boring, loutish, anti-social, changeless, risk averse people.* Law enforcement is now their specialty. They are like a breeding bureaucracy of government. Executives come and go but the HR Poddies stay on for their next kill or at least the slowing down of progress.

They actually have a fondness for other Poddies in the company—the do-nothing Poddies of various other departments, because like them, they are thought to be harmless people who would never take a chance. People who are risk-takers and are entrepreneurial within a company are dangerous so far as the HR Poddies are concerned, because out-of-the-box thinkers like to defy regulations.

So if you are an out-of-the box thinker, have ideas and convictions be sure to watch out for the HR Poddies. They're out to pass fluid on your parade. If you do not fit into a certain box or criterion or aren't compliant, you will be deeply under suspicion and could be on your way to HR hell, for they are the most nefarious Poddies of all. For they actually do something and it ain't good.

Of course, right behind them in a corporation or institution are the staff lawyers whose main job is to protect against the prospect that new products and services—anything new period—should find their way out of the company. The lawyers usually detest the marketing and sales people who are assertive with new ideas that might provide company growth and add to profits. But these are unproven ideas, products and services and they might cause the company to be sued if something went wrong somewhere or the products and services proved to be failures.

Funny thing is all companies were small once, all of them. You usually start with one person or several others, working out of a garage or a small warehouse and your creative juices fly. There is no attorney in the next room to "help" you or an HR person standing wing ready to tell about you the essence of life. It is a shame that when these little companies become successful they often lose what made them happen in the first place—an esprit de corps, an entrepreneurial drive and a willingness to take a calculated risk.

But by not taking a risk, these companies are actually putting themselves at true risk's door. Not moving with the challenges of the marketplace and competition can make them fallow and affect the top line as well as the bottom line. Not to say make the company stuffy and boring. It is ironic that success can often breed inertia. That second generations of a company don't have that certain je ne sais quoi. Maybe its because they get mired in regulation and Poddie mentality. Or, they don't have the killer product developers and sales people that the original generation had.

Some last longer. It took IBM three generations to get sick, and sick it was, caught totally off guard in the personal computer market. Likewise, it took the amazing franchise of GM several generations to get bland and customer repulsive, while the Japanese automakers went to town all over America with exciting, stylish and well working machines.

But there are some good exceptions. Even very old companies can become entrepreneurial with the right leadership. When Invacare was 100 years old it was dying, only a few million dollars in sales and going no where. Then Mal Mixon bought it from a parent company that didn't have much use for this loser. In the next 25 years he proceeded to make it a viable wheel chair manufacturer and healthcare-related accessories products company, and he propelled it into the billions in sales league. From the oldest and smallest company in its field, it has become number one worldwide—through ingenuity of product, including sports wheelchairs and many other products beyond the basic wheelchairs, employee incentives, and a general colorful approach to doing business.

—Ron Watt

Somehow, the good companies new or old figure out a way to overcome the inherit obstacles that every organization faces. They do it with common sense and good leadership—and creativity—in spite of those possibly within their own organizations who can't or won't see the big picture.

In an exceptional company probably 10 percent of its employees are truly outstanding. They come in early, turn on the lights for the rest, they make outstanding things happen all day and they turn off the lights after all have left. In a truly outstanding company, the CEO spends a lot of time trying to make it 15 percent. The five percent gain in worker quality could be mind boggling in terms of company returns. Revenues, profits, morale and new product innovation.

Good companies put a premium on product and service development, on plenty of time with their customers (in person or <u>live</u> on the phone, not just through e-mail), and on encouraging employees who detest being Poddies to march forward to help with success and profit. And build on the new ideas that make a company great.

There *is* a way for Poddies to escape. More about that later.

Sidenote to H.R. Department:

"Melancholy is the pleasure of being sad."

—Victor Hugo, with a thanks from Dr. Mardy Grothe's Oxymoronica (Harper Collins)

The size of the HR Department typically is in inverse proportion to the company's growth stream. The larger the department, with its tentacles casting every which way, the less the stream, the less the steam. Companies that scale back on sales people and add to HR or other internal service departments are usually heading in the direction of immutable demise. They are shackled from creativity and innovation, handcuffed by over-interpretation of regulations and mired in a muck so palpable that it would make the sci-fi fright film *The Invasion of the Body Snatchers* seem tame by comparison.

"We are neither hunters nor gatherers. We are accountants."

CHAPTER SIX

**Watch out for the bean counter
Poddies as well.**

IT IS POSSIBLE THAT THE ORIGINAL PODDIES in corporations might have been the bean counters, now known as chief financial officers, controllers, comptrollers and their staffs. They're refrain is: "show me the numbers, show me the business case, why should we do this?" They never ask about market share or growth. Just numbers, numbers, numbers!!

These are often people who look askance at sales, marketing, brand, advertising and PR people—their customer care counterparts in the organization—and always think that too much money is being spent on these activities, never understanding the human dynamics that are involved. An old saw says "that without good customer service and products, there aren't any beans to count."

The bean counters seldom understand human dynamics and don't want to. They don't understand branding and marketing initiatives. They think the best way to control customer care costs is to have them regulated—*by bean counters.*

The bean counters and their like are often self-appointed experts on all matters of the company. Like bankers, they are extremely strong at looking backward without any ability to look forward—vision. The chief asset of leadership is always vision.

They would rather succeed thinking small than risk failing thinking big. Their predecessors wore green visors; maybe they should as well, and maybe blinders, too, like an errant racehorse.

When they think of downsizing, they always think that the company must start with the sales staff. Almost never the financial staff, IT or HR. They look a downcast eye at sales, never considering that it is the salespeople who bring in the customers, and it is the customers who provide the revenue that pays the financial department's salaries.

They are remindful of NBA and NFL players who take a nasty look at the fans—their customers—and do all that they can do to avoid signing autographs or posing for photos, never remembering where the big money to pay their salaries comes from.

No company can exist without strong financial control, even small ones, but when the financial cart gets ahead of the pony a Poddie condition starts to exist and it is never good.

I have hardly ever known a mathematician who was capable of reasoning.

—Plato, B.C., The Republic, from Leonard Roy Frank's Quotationary (Random House Webster's)

Forward-thinking individuals in companies must learn to play Rocky with these financial Apollo Creeds. Tell 'em where to go if they get too high and mighty. Remind them how their salaries are derived—and it isn't from yards and bushel baskets of financial data—it's from revenues begotten from sales, branding and marketing. Tell them to help you protect the bottom line but also tell them not to mess with that which creates revenue growth. You don't create revenue growth by adding to the HR staff or the financial department, you create it with good old sole-to-pavement people who consistently bring in more business than they cost. Pretty simple.

A company becomes asinine, and surely Pod-ridden, when it takes away people who are the producers and adds to departments where process is king. People who make things happen—through sales—are a gift to the company, not a detriment. People who sit around working their machines and their processes are the ones you should be watching most, for they can drain a company in more ways than one. And they rarely produce revenue for the company.

The financial officer and his staff who see the big picture and honestly understand where the money they count is coming from are the rarity. And this has been so for just about time immemorial. That's why they're called bean counters.

Often in these days of severe regulation and general angst, financial officers—numbers nerds—are wending their way to the top of corporations. They are there to make the company tow the mark on profit making. Significantly, the bottom line each quarter and each year is paramount. Nothing else much matters. Our whole business society is driven that way more and more.

We find it, however, disconcerting that while marketing and sales executives who make it to the CEO's suite are without question required to have a good handle on the company's financial matters, a real solid knowledge of the balance sheet, cash flow, revenues and operating profits, the reverse is not often true with financial people. CEOs who come from finance quite frequently have a shabby knowledge of how the customer care side of the business works. They are left lacking in their knowledge of marketing, branding and sales because no one ever forced them to learn those segments of the business. In today's world, most customer transactions are treated just like an Internet sale. No face required, just a credit card. They typically totally grew up on the bean counter side.

Now as CEOs they are forced to learn on the job. For many, the timing is too late, and ultimately the company suffers for lack of external focus.

Smart companies, like GE, Hallmark, Honda of America, American shoemaker Allen-Edmonds, and paint giant Sherwin-Williams, force their budding executives to work in many departments and operating divisions of the company, so they get a good handle on the way the entire organization works. A year there, two years here and other ongoing, pertinent, solid experience within the operations make for a solid chief executive when his or her time comes. In this method, the chief executive has the knowledge of the strengths and weaknesses and has the background to generate the ideas that can push the company forward during his or her leadership.

Certainly, no CEO should ever reach his or her seat without a fundamental sense of customer care and on-the-ground sales. For the CEO must always also be the Chief Sales Officer. He or she should always be the antithesis of the Chief Pod Officer, who is always:

Weak, complacent, tedious, predictable, boring, loutish, anti-social, changeless and risk averse.

"He's charged with expressing contempt for data-processing."

CHAPTER SEVEN

Here's a whole city of Pod Heads . . . IT
Command and Control.

THE MODERN-DAY PODDIE could not be better depicted than in the embodiment of the Information Technologist. This specialist is a cartoon-like character who mostly lives best in a cocoon-like labyrinth with others like him or her. Going out in public rains terror upon IT people and their only counter punch is to talk a gumbo of Pidgin Sanskrit that only they know.

Deadlines are beyond them. If they mumble something about taking four months, you can be sure it will be longer than a year. Massive cost over-runs are often their by-product.

Ask them one simple question, and it will turn into ten conflicting answers. But power they have because clearly almost nobody outside of their Pod understands what all these electronic gadgets do and how they work. The more you try to find out, the more theater of the bizarre it becomes, to the point that you think it is better to invest in a fine writing instrument from Waterman or Mont Blanc. The only bad thing *these* gadgets do is leak ink every so often.

These are strange and different people, indeed. But the curse of it all is we can't do without them and that is where these bespectacled little despots derive their empowerment.

Just like the HR Pod Heads, they are sorely needed but not loved.

Information Technology is the "Command and Control Center" in most businesses today. This Poddieville is not located in the heart of operations, though. Quite the opposite. But does it ever have its impact on virtually everything we do.

IT Poddieville may seem serene but clones they are a mak'n, programs they are a writ'in, processes that you must follow they are documen'n and their reach captures everyone from payroll to profit centers. There is only one way to do it, and "it's our way, and everyone must do it the same!" To operations people, that is "Command and Control."

The ultimate buzzword that has come up, down or sideways from "Command and Control" is "paperless." That should stop you right in your tracks! A continuous, unrelenting brainwash campaign now has every manager overjoyed with the final word from Poddieville. No more file cabinets, no more clutter on desks or no more paper trail to fool with. This IT Poddieville is so reliable, why worry?

"All the information that you will ever need, we will keep for you in our electronic database at 'Command and Control'." When you have a need to know, "Command and Control" will give you a security code to get into files that you are authorized to see. Don't worry, our systems are almost always up and ready for your inquiries and work processes.

TRUST US, YOU WILL NEVER NEED A PAPER TRAIL AGAIN! COMMAND AND CONTROL IS WATCHING OVER YOU . . . WE WILL DO THE WORRYING FOR YOU.

The vast number of Poddies that occupy "Command and Control" look really non-threatening, don't they? Watch what programmers, analysts, user interface managers or facilitators do. Watch them get to their Pod, flip on the silver screen, check their BlackBerry for orders from the big Poddie, and strap on the cell phone, waist high and at a convenient quick-draw range. That is what "Command and Control" looks like in organizations today. These are the modern gunslingers who keep their noses on the barrel— doing the same things every day and every week. "Return of the Body Snatchers," tell us it is science fiction, please.

If there ever were an "enforcement patrol officer" or "Poddie Cop" for the Body Snatchers, it would have to be these rapidly multiplying people in our ever-growing Information Technology departments. These departments are endemic—or shall we say pandemic —to companies of medium and large size today. For people who love the practice of "customer care" inside companies, this department and its minions have to be the ultimate definition of *People of Dullness and Drudgery Imbedded Everywhere* they can get in your way. IT Poddies represent the very body and soul of body snatching.

A delightful recent book, *Fish*, is a primer on how to take these jobs of drudgery and turn them into excitement for the workers. But we're not sure even *Fish's* concepts of fun, energy and positivity would work down in the bowels of IT. Down there somewhere is the Victory Boulevard of Pod life, cubicle after cubicle, row after row. And in the town square down the boulevard is a veritable giant hornet's nest of Big Trouble.

Programmed languages are IT Poddies' daily fare and their output demands that everyone using their magical solutions have to apply them in exactly the same way. One failure on your device and a call to the "help desk" at Poddieville is the only solution. Now the real trouble has just begun. They will return you to the "clone track."

These departments are the embodiment of IBM and Microsoft and all the other hardware and software companies that have financed the "The Return of the Body Snatchers." Billed as the panacea for cost savings, efficiency, penetrating research and development, and distinctive value added adornments to the marketplace, these "enforcers" have produced generic solutions that are costly and time consuming. The common result is that very little value is added to products and services and, instead, we have a cornucopia of complex confusion for the customer, the sales staff and everyone else involved.

But the notion continues that transactionally based companies cannot do without these machines' throughput in the service of their customers. And that is so true. But at the same time, the body snatchers have managed to put every employee behind a screen every day, most every hour or parts therein. For us, there is no way out except to the operating room to take care of their Carpal tunnel syndrome or to the Lasik doctor to fix our failing eyesight.

The "enforcers" are the lead dogs in making us *weak, complacent, tedious, predictable, boring, loutish, anti-social, changeless and risk averse.*

What a syndrome of doom.

The "enforcers" have unique and subtle ways to graduate us to the Pod, to clone us as look-a-likes. Sign-on's, coded entries, secret languages, secure sites, domains, software licenses, e-mail trails, all-inclusive files, private files, and on and on it goes, to allow them to keep track of our every move to make sure we follow their book, their procedures, their processes. Using their devices is certainly not the place to get innovative or inventive. Ideas that break out of their book "can be really expensive," they say.

Today, Information Technology Centers are all powerful. This from people who can't communicate in normal human ways. Truly amazing.

As their powers grow and they create more clones, company after company and employee after employee, they have also become internal branding specialists. About every five years or so, they change their name to better reflect their growing power base. As neophytes in the company years ago they were known as "data processors." Not a powerful name if you think about it, but a start! Data, after all, is simply data, and certainly was not the type of thing that you could thrust on an entire organization and capture every employee's soul.

As their aura and power began to grow, they renamed themselves "information technology," a more powerful word that suggested there was actually some good, useable information in all of these data elements (reams and reams of paper when printed out). Now that they have permeated every corner of the company, and captured and cloned the masses, the all-powerful, dependency label has just been introduced, an oxymoron . . . "The Business Solutions Center!"

In reality they should simply be one of the tools that the organization and management use to move into or protect a leadership position in their industry. They provide files where data can be entered and retrieved in a somewhat informative way. The information provided becomes a "tool" amongst many options we have available to solve problems, create products or to provide customer services. They enable us to print documents in a rapid and reliable way and they allow us to build and retrieve history about people and transactions. But this is not: "The Business Solutions Center."

Why do highly intelligent, highly paid people want to do this to us and our companies? Why have they inflated themselves beyond their true purposes? What's the reason for this unforgiving power trip?

They so want to possess us with their technology— almost a sci-fi thing, but unfortunately it's real, folks. Why do they believe that everyone must act alike and do things alike to progress in their jobs? Do they believe that typing is a highly sought-after skill? Just because they are device drones, must we be one as well? Does the new Information Technology, the Business Solutions Center, put us on the be-all, end-all highway to success or have the body snatchers sold us a bill of goods so that we will be Poddies like them.

In today's world, Jerry McGuire proves to be wrong: "Don't follow the money, follow the Pod."

The horror of horrors will be when we began seeing droves and droves of CEOs coming out of IT—The Business Solutions Center. People who don't know how to sell, communicate, understand finance, don't appreciate marketing, branding, advertising and public relations, but just do "their thing" as they have always done from their Pods. Let's just make the whole corporation one big Pod.

If there were a new movie, "The Return of the Body Snatchers," I would surely warn you that information technologists and their device delivery systems are possessing us. We are being put into these cubicles with screens and all the other electronic gizmos, just like the Poddies have. Those comfortable, non-threatening Poddies. We are being cloned by these new Body Snatchers into the walking dead—device drones without emotion, drive or charisma. New and potential customers crave attention and like to bond with interesting, innovative people—customer care people. And this is the bunch—the IT Poddies—that we too often ask to care after them? Let's give those customers another dose of weak, complacent, tedious, predictable, boring, loutish, anti-social, changeless and risk averse. They're getting used to it. Too bad.

—Dr. Miles Bennell, protagonist, Invasion of the Body Snatchers, RIP . . . "I told you what would happen."

Go with your heart!

CHAPTER EIGHT

Are you a Podophobic? Good.

IF YOU TREASURE CHANGE, you probably are a—eeh gads—Podophobe. The opposite of a Poddie.

Pods make you nervous, and rightfully so. You worry that you might succumb and become one of them. But rest happily, you are on the road to success that the acceptance and actualization of change will bring.

Sometimes, people (probably Poddies) refer to a "Cowboy," "Swashbuckler," Musketeer." Take those words a compliment. The Poddies are afraid of you but you don't need to fear them, because they would just like you to go away, get out of their space, and keep marching past them.

When you do have to encounter them, they might say something patronizing like, "Why don't you stop and smell the roses?" (Read: "Why don't you become lazy and conformist?" like us!) "Why are you a saber-rattler? It's better to be like us. Your emotions will destroy you. Be like us."

Be a clone drone. Be self-satisfied. Be smug. Don't ever question yourself. Don't question anything. You'll make a lot of money (or so they think), you'll have new SUVs, a McMansion and you'll be the life (read: stiff) at neighborhood parties, talking of your stock portfolios (read: small) and where you are sending your clone kids to school (read: let's all be the same).

"You start taking a lot of chances and you'll be on the scrap heap before long (read: don't be an entrepreneur). And you'll be sorry."

Let's run in packs, let's bore the hell out of each other and let's brag about our accomplishments in Poddieville. Let's live with the rest of the Poddies. Let's ingrain in our cocoons. Let's have the same kinds of Poddie dogs as well as kids. Let's not suffer rebels kindly. Let's conform. Let's bring about nothing new, nothing different. Let's all watch the same movies, read the same books. Let's have the same mail boxes in front of our McMansions. Let's have the same grass mowed by the same guys who also put the same insecticides and herbicides on our lawns. Let's have three-bay garages and even the same garbage cans behind them. Let's have the same cedar decks. Let's question anyone who resides outside our comfortable, exact compounds. And, finally, let's even get divorced and recycle another family of clones.

Let's be pseudo-intellectuals, avoiding the rest of the colorful world. Let's wear the same clothes—bow ties would make us suspect—and let's eat in the same restaurants and talk incessantly about what we ate at those same restaurants when we run into each other at soccer.

Say, Poddies: Why don't you read Voltaire? Something like *Candide*. He'll tell you more about life today from his perspective of 400 years ago than John Grisham and Scott Turow do. He'll tell you what the word "banal" means and maybe it applies to you. But you are too cloned and unaccepting to be bothered with his drivel. Of course, you've never heard of him or don't want to. You are a mercenary of monotony and you like it that way. You are out to make your organization *weak, complacent, tedious, predictable, boring, loutish, anti-social, changeless and risk averse.*

I was at a cocktail party in Cleveland Heights shortly after LeBron James became the first draft pick in the NBA and was selected by the Cleveland Cavaliers. At the same time LeBron signed a contract for Nike for nearly $100 million and other contracts with other sponsors for millions more. A woman said that it was detestable for a young man of 18 to set such a bad example for other young kids. In other words, he should have gone to college first. He should not have gone to the NBA directly from high school.

I said to her he had more than won the lottery with his stupendous talent, good timing and skills as a kid mature beyond his years. Going to college would have been nice. But what if he got hurt, what if the $100 million and more in sponsorships, the big NBA salary and many more promotions that would bring him even more millions never occurred? How smart would

LeBron James be to not take advantage of opportunity when it knocked?

The woman was adamant that he had done the wrong thing and would lead other students to bail out on college thinking they could be the next LeBron James. She was a senior counselor in a private high school and took umbrage with LeBron for his decision, overlooking his unique abilities and place in time.

I told her I felt the same about some young kids whose garage band suddenly gets discovered and they "make it," are on the map in rock and roll or rap or hip hop. What should they do? Forget the opportunity, go to college for four or more years and then try to get it re-started when they "have their education." I said they should go for it. Few people—including those college educated—might ever have the opportunity to have that kind of success doing what they love doing.

But the counselor didn't buy that one either.

She couldn't.

She was a clone.

—Ron Watt

Sometimes it is all right to just go with your heart, and see what happens. "LeBron To Start All-Star Game In Sophomore Season" screamed headlines in *USA Today* and other newspapers. *Sports Illustrated* had him on its cover with the words: "Best Ever?"

The spoils seem to go to those who are unique, innovative, tenacious, and risk-taking. Talent is always important, too, but talent can often be overridden when complacency takes hold, when the person is playing "the game" overly carefully and close to the vest.

Invention and creativity can come at any age in a person's life.

Much of titan architect Frank Lloyd Wright's success came after he was 60. So did Pablo Picasso's, Ronald Reagan's and Mother Teresa's.

Many other well known people became famous and successful in their fields later in life. One recent study of 400 celebrated people indicated that well over 60 percent of them gained their success after they had reached their 60s. Some even were in their 80s when they gained their celebrity status and greatest success.

There ain't no rules around here! We're trying to accomplish something!

—Thomas Edison, from "The 2,548 Best Things Anybody Ever Said," compiled by Robert Byrne (A Fireside Book—Simon & Schuster)

THE ROBOTIZATION OF AMERICAN MANAGEMENT

CHAPTER NINE

The paralyzing fear of change.

THE MOST DESTRUCTIVE and debilitating human characteristic for any organization today is people who have a fear of change, or put in other words, a comfort with business as usual. This must be part of the creed of many Poddies in an organization, in their cubes, comfortable and content with good salaries, adequate bonus plans and benefits. Their personal lives could not be a whole lot better. Amazingly, over time these once energetic employees become blind to the dynamics of change that is swirling around them personally and in their business. Of course, the marketplace change agents are: demanding customers; politicians enacting new governmental regulations; competitors trying to get an upper hand; and stakeholders who are demanding higher performance.

The cubicle dwellers that we call Poddies don't like disruption, very content to live out their working lives with business as usual, especially if the company continues to perform in a satisfactory manner. After all, why change when everything is going well?

———

Does this sound like a "Poddie Plan" that was followed by the employees and management of the major air carriers in the United States? Lack of dramatic change at key times in their business history has led to their ultimate demise in many cases and at least a bad rattling of the cage in others. History does repeat itself as earlier we had witnessed the very same thing in the railroad industry—turning a once grand king of transportation into a weak skeleton of its former self. Things were going well for the airlines just a couple of decades ago. Airfares were high, air travel was growing and some airlines were actually making profits.

Dramatic, history-making change occurred in 1978 when Congress passed legislation deregulating the airline industry. Suddenly the F.A.A. no longer would supply the security blanket for major carriers to raise rates year after year, watch quietly as union negotiated wages continued to rise rapidly and then simply wink when world-class benefit plans were negotiated by pilots, flight attendants and ground and maintenance crew people. Under government supervision, the airlines had managed to build an infrastructure of expense and future variable costs that would expose them to ultimate destruction in their new climate of "free enterprise."

Labor unions and airline management are one of the best modern examples of the cloning of practically every aspect of the airline workforce. Contracts read alike, salaries and benefits mirror each other, and behavior is in lockstep, a perfect example of Podddies and the world they live in. Now havoc is wreaked throughout the industry as the few airlines left try to hold on and the unions are scrambling to survive as well with givebacks they never imagined would ever happen.

Where were the management and union visionaries in the 1990s? Where were the change agents when they were needed? Instead of change when it was needed for survival, airline executives elected to introduce frequent flyer programs and other gimmicks in an effort to protect and enhance their market share, rather than take on the daunting task of reopening union contracts or playing hardball at the next contract period. Suddenly, thrown into a world of unregulated competition, was there not one bright airline executive or some entrepreneurial, leading edge employees or union that thought to step outside their offices or hangars and look to the *Southwest, Southwest, Southwest?*

There must have been Poddies imbedded everywhere in the industry. Certainly, everyone that was cloned under a union contract was not going to suggest change. If big change is coming, do not ask us to change or we will strike. The lack of flexibility and ingenuity and the fear of change finally has struck and struck hard. The 9/11 tragedy might have sped up judgment day but the fate of the major airlines was sealed by their lack of creative action following deregulation of the industry.

We constantly talk about hindsight being 20/20, and true to that aphorism, fear of change, union pressures and/or poor management talent and vision sealed the fate of this industry in the 1990s. The lack of fundamental operating changes when they were needed turned out to be deadly for the most basic of passenger transportation businesses. It just didn't seem possible, did it, when things were going well?

A fleet of do-nothing, business as usual Poddies can bring superstar corporations to a grinding halt. Poddies view risk-taking and leadership initiatives as foregone failures and they are not afraid to contribute to those failures by their inactions when instead positive actions are desperately needed.

The single biggest contributor to the failure of new products, new services, expansions, acquisitions or diversification efforts by a corporation is certainly not a market mismatch, poor due diligence or market timing. Most of the failures are self inflicted by a bunch of disinterested Poddies that are more interested in "business as usual" than new successes that require inspired work, extra hours and innovative thought. The whole process of innovation, they think, will probably change their work habits, change their job and change everything they consider sacred in the workplace. *"No way!"* they say. *We like weak, complacent, tedious, predictable, boring, loutish, anti-social, changeless and risk averse.* That is their mantra.

Change is so dynamic in the workplace today that most leadership companies monitor employee and marketplace results monthly if not weekly. It would not be extraordinary for leaders to make competitive course changes quarterly or at least in a six-month period of an annual plan. These course changes also require course changes by employees, sometimes twice a year or more.

The undoing of many fine companies will follow the airlines, if we continue to allow employees and executives to isolate themselves in cubicles, behind screens all day with their devices. In this way, they lose forever the "feel" for the dynamics of change and successes that would come simply by embracing and taking advantage of change as many times as possible.

Instead, they should be embracing change and taking advantage of change as many times as possible. Being an agent of change is certainly always uncomfortable, but now an absolute in today's tiny competitive world. To do so may require more work, but we assure you it is more fun.

In an earlier chapter, we talked about the 10 percent of employees in most corporations who are truly outstanding. They make good things happen for CEOs who are change agents. But they are abhorred by the Poddies.

These 10 percent are spread too thin in most companies Most often they are the busiest people around the office. They are members of every committee or task force and they are systemic risk takers. They are the Bogeyman of Poddieville . . . Thank God.

For corporate executives in leadership companies there are two great rewards involving employees. The first is giving young people some rope and seeing how they respond. Most of these corporate upstarts have no fears, no fear of change, no fear of failure and no fear of hard work and long hours. The challenge is to keep the fires burning for long periods of time, and certainly do not give them any guided tours of Poddieville. Poddies can be found in the bowels of most any corporation in various departments such as IT, Financial, Legal, HR, Operations and elsewhere. They are slow and stifling turn-offs.

The key to handling the magic 10 percent is job rotation, ego recognition and keeping them out of the "class system." These sorts of creative executives can get bored easily; keep 'em busy and diverse.

The second reward is seeing the escape of veteran employees who are challenged by the thought of life outside of the Pod. These employees are typically nine to five people, good people, who want to become part of the magic 10 percent. They are frustrated by being in the great unwashed of 90 percent and want to do something about it. They are willing to work harder and for longer hours. They want to emerge out of the Pod and can see life is better in the rarified air where things are happening, things are changing and more exciting. They might not aspire to key leadership positions because they do not like the pressure—certainly they have fought change in the past—but with the right opportunity they just might ascend out of the Pod and become key contributors to the company.

The boredom, lack of challenge, or maybe the lack of outside business contacts just gets to a few of the veterans. They start to ask for new assignments; they might even want to do some "field work." Management is shocked to hear a question like "what do we have on the burner that I can help with?"

Finally, they have tired of sending e-mails and voicemails all day, or entering meetings and reviewing things to do on their Palms and BlackBerrys and Axims. Suddenly, they have thought to pull out a copy of "the annual plan" and "the mission" and can see a role for themselves to help move the company ahead. Maybe 10 or 15 years behind the screen in Poddieville are not what things were cracked to be, they begin to think.

The escape of these veterans is the more rewarding of the two scenarios because these escapees become disciples of the plan and mission when challenged and rewarded. They have the knowledge of the culture, the politics and chances of new and sustaining successes for the company and themselves. Suddenly, they are mentors to young starters, mentors to the sales and marketing people and they become risk takers along side of management. They are the corporate born-agains. Escapees are the best!!

"This is so cool! I'm flying this thing completely on my Palm pilot!"

CUBICLE POOLING

CHAPTER TEN

How did I get conditioned to be comfortable with Pod life?

FOR ONE, YOU ARE MORE AND MORE risk averse. You don't like risk-taking people surround you. You think that you will lose your job if you take on any aspects of risk/reward. Your reward is comfort within your Pod, and you don't want to change that.

But risk-taking is often confused with gambling in Vegas or some other place with casinos. The truth is that risk-taking is the exact opposite of gambling, where the odds are with the dealer, not with you.

Corporate risk-taking odds can be increased immensely in your favor with good planning and outstanding management people, a simple formula that is difficult to achieve in a Pod-prone organization. This formula gets easier in an organization that has top management that is not complacent.

The biggest risk in an organization is promoting the wrong people and not advancing the right ones—

the people who are willing to take chances and meet opportunity square in the eye. But in organizations, in spite of themselves, Pod people rise sometimes to new levels to make their company *weak, complacent, tedious, predictable, boring, loutish, anti-social, changeless and risk averse.*

You can tell who they are because they question every move in every direction. They put down progress over process. They like to form study groups with long-term deadlines and often what they study over extended time goes nowhere. They make their salaries and bonuses, but what do they really accomplish for their corporations other than impede progress? If you allow them to, they will gather together in their comfortable groups and do everything they can to block the risk-takers who might take the company to another level in R&D, product and service branding, and new flows of revenue. Their goal is to prove how it won't work rather than how to make it work.

Could Hamlet have been written by a committee, or the Mona Lisa painted by a club? Could the New Testament have been composed as a conference report? Creative ideas do not spring from groups. They spring from individuals.

—A. Whitney Griswold, Yale University baccalaureate address, 9 June 1957 (thanks to Quotationary, edited by Leonard Roy Frank—Random House Webster's)

The sadness and the humor of it all is these same people never come up with other ideas that might help the company. They are too busy proving what won't work, staying stolid on the successes of the past,

thinking that these will carry the company forward.
This philosophy seldom works.

*One of the toughest decisions that I had to propose
to our board when I was CEO was to sell our life
insurance company at the pinnacle of its profitability.
It was becoming clear to me that the investment in
new product development and marketing costs were
going to be prohibitive as we encountered the need for
an "equity kicker" on our products going forward. So,
we sold the life insurance company for a near record
multiple of earnings and reinvested in our core property-
casualty-surety business as well as in several other new
financial services businesses. We also acquired a small
property-casualty insurance company that proved to
better serve the overall mission of the group.*

*Meanwhile, our merchant bank was beginning
to get off the ground in several states, title insurance
was becoming part of our portfolio as a result of an
acquisition and several other businesses that could
serve the financial services sector were started or
acquired. This helped to balance out the cyclical nature
of our core businesses. As in the case with most small or
start-up businesses, a couple did not work out and were
jettisoned before the red ink got too bad. Developing
an entrepreneurial attitude about these new businesses
was a difficult task not only for me, but also for our
board and the other executive officers. "Business as
usual" would have been a lot more comfortable, safe
and secure.*

*Calculated risk-taking in a market-leading company
is always necessary to stay at the top. Standing still is
not an option! At times, companies need to be added*

or deleted, missions changed to reflect a dynamic marketplace, and the scope and nature of your business must be reinterpreted. Your biggest asset, your employees, must "get on the bus," get energized and chauffeur the company to new successes.

I enjoyed being the "change agent" for our group of companies!

—Cary Blair

*   *   *

Any new venture goes through the following stages: enthusiasm, complication, disillusionment, search for the guilty, punishment of the innocent, and decoration of those who did nothing.

—Unknown, from "The 2,548 Best Things Anybody Ever Said," compiled by Robert Byrne (A Fireside Book—Simon & Schuster)

Poddies, or People of Dullness and Drudgery Imbedded Everywhere, as we know them, don't have the facileness to effect change. Inertia is their game and they like it that way. The best they can do is find out what other people—other companies—are doing and try to do the same thing, because they think it works or it is the safe way.

This makes us think of the National Football League. Is there anything more Podified than this? Each team looks the same, acts the same, and in reality is the same. There are always ingenious exceptions, the work of geniuses in a miasma of sameness.

One has to only look at the success of New England's Bill Belichick and that of Bill Parcells, now of Dallas. Both of these guys have won Super Bowls, and with different teams, and still they persevere, beating the NFL system that offers very little creativity from team to team.

Most of the rest of the players and coaches just get recycled from team to team. They are interchangeable and exceedingly boresome. Of course, all the teams are highly computerized and the computers tell them how to act and react against their competition. In other words, how to act and react against themselves, with the same size and speed of player according to position. For example, every offensive lineman must be over 300 pounds, every defensive lineman under 300 pounds but fast. Quarterbacks are best 6-4, tight ends 6-6, 270 pounds, wide receivers 6-2, 200 pounds, defensive backs 5-10, 185 pounds, etc. Is it any wonder that so many of these NFL games are so slumber producing?

One coach told us the computer analysis is so hot that he could find a one-handed, left-handed defensive back who had only been in jail six months and now was residing in his mother's work shed in Tacoma, but has 4.33 speed. The coach said he could find this fellow in a matter of minutes and bring the player to his team the next day if another player were injured. We only ask, does the computer measure heart as well as data?

Likewise the NBA, where the only part of the game you really have to watch comes in the last two minutes. The rest of the game is a bunch of big guys and four little guys incessantly running up and down the court trading shots. Our thought has been that the fans would be better served to watch these teams play six two-minute games a night for about five weeks straight and call it a season. Why kill us with so much sameness over such a protracted nine or 10 months? Let's get some spark going.

Now baseball is a different matter. Baseball players come in all shapes and sizes, and the game is played the same way it was 100 years ago. But there is a nice symmetry here. It can lull you to sleep or make you go crazy. Nobody knows what's going to happen or when it's going to happen. It is predictably unpredictable from spring training to the World Series. And the beer, hotdogs and popcorn just seem to smell and taste better than in the NFL or NBA.

But mostly, sports at the pro and collegiate level have seen the cloning of most of the coaches and players—just like any other business today, more and more managers and employees coming out of the same Pod. Is there no end?

The man who lets himself be bored is even more contemptible than the bore.

—Samuel Butler, The Fair Haven, 1873, as compiled in Quotationary, edited by Leonard Roy Frank (Random House Webster's)

"When it comes to loving someone, I never seem to get it right."

CHAPTER ELEVEN

Ever eroding people skills.

ONE OF THE MOST DISASTROUS RESULTS of the Body Snatchers is their robbery of people skills from our young adults who aspire to be executives and, to some degree, even our sales and marketing veterans, who should know better. In most cases, these skills are vital to personal success and to the organizations for whom we work for pay or as charitable volunteers.

As the Snatchers brainwash us daily with the attributes of new devices, they also rob us of the crucial skills necessary to get along with people and to influence them in certain ways that can only be accomplished face-to-face.

E-mail is simply not an acceptable substitute for a hand-written thank you note . . . or a pleasant "live" phone call or lunch out. Most of the e-mails you receive today are misspelled or are horrible examples of acceptable grammar. Sometimes all you read is gibberish—something dashed off with no feeling or passion for the tasks at hand or the product or service being sold. At best, e-mail is being used for "yes" and "no" answers, or worse, it is inordinately long and convoluted and difficult to read and understand. This is what the Body Snatchers are doing to us—in the name of efficiency and speed.

Good customer care people should become big fans of Peggy Post and Peter Post, famous names in the world of etiquette—and now especially in business etiquette. Their book, *The Etiquette Advantage In Business*, now ironically serves as one example of key customer care skills that are not being learned by most Poddies, but *must* be acquired for those who wish to escape Pod life and be successful in customer care.

Those who wish to spend their lives out of the Pod must understand and practice these social skills. They must understand how such skills, or the lack thereof, can cast a positive or negative reflection of their organization. With customers and prospects, the impression we make, even initially, is always long lasting.

If we are to dwell outside the Pod and have escaped from the screens and keyboards as much as we can, we are going to be dealing with—people! Many of the people we'll meet are important in their lines of work or charity. We will engage them in many ways. We might be asked to give a speech or teach classes. We may be invited to black tie dinners with four, five or six courses of food and wine. We may meet ranking government officials or foreign dignitaries where proper greetings and introductions at times may be more important than the subject matter. We may be required to wear business attire, not casual wear common to Pod life. We might be asked to select a good wine for an important sales dinner.

The Return of the Body Snatchers has robbed our young people of these skills and propriety, because assets simply are not compatible with the device world. We cannot spend life behind screens and keyboards and learn the important social graces required for customer care.

Proper social conduct, good manners, proper courtesies, dress codes and other common social behaviors have been lost on Pod people. It's difficult to learn how to handle a dinner when most of one's dining time is spent at fast food restaurants or cafeterias—where everything is taken back to the Pod. Not much finery learned there.

We should never take for granted what might happen in social situations. We have to be quick on our feet and be sensitive enough as to how to roll with the flow in social interaction. It is better to be prepared, and it just takes a little more effort to do so. People skills make big differences, even in today's crazily fast-paced world. Bad social skills will never be forgotten.

When I was CEO there was a local politician who was one of my "best friends" when she was trying to attract a contribution or sway opinion her way on a given issue. I also remember that in other social situations not related to her, I was barely recognized or acknowledged by her. Despite her favorable party affiliation in my eyes, I no longer vote for her. Consistent customer care is a good thing.

—Cary Blair

We consider the Post book our security manual for customer contacts. The world has become small, contact with people from other places and countries is quite common, mores are different from one place to another, proper conduct in social situations is more important, not less.

My last inquiry into my Peggy Post Security Manual involved a dinner that my wife and I hosted for the ambassadors to the U.S. from Egypt and Mexico, accompanied by their wives. I would be introducing them to the other guests during the evening. My inquiry was to the introduction of the spouses—to find out the proper way that was first respectful to the ambassadors and at the same time sensitive to the presence of their wives. Ergo, His Excellency Anwar Abar, Ambassador of Eygpt and Mrs. Abar, the Honorable Jose Hyak, Ambassador of Mexico and Mrs. Hyak.

—Cary Blair

In a short period of time, less than five years, electronic devices have created the most inconsiderate population to inhabit the earth. We can see them with their gadgets everywhere we go: on airplanes, all over airports, on the sidewalk, streets, in restaurants, at the grocery store, in private meetings, in movie theaters, concerts, and, eeh gad, even at the symphony and the opera.

Our thoughts especially go to cell phone technology and what this has done to intelligent people. It's unusual today *not* to encounter any person eight to 80 who does not have a cell. Everyone must have one and the rational for just why can be interesting. It has gotten so bad that most fixed line pay phones in public places are non-existent today, for lack of use.

What if we had to give up one half of our phone time by law? What if we could not take our cells into a public place—period? We could only use them in a private setting where no other person could hear our "private conversations."

Let's face facts: These damnable devices have turned most all of us into social jerks, when before most of us were nice, considerate people.

"Are we thinking here, or is this just so much pointing and clicking?"

CHAPTER TWELVE

Where the "partnership" with technology goes wrong.

THE RETURN OF THE BODY SNATCHERS pokes a lot of fun at the device world and the infatuation many people have with the new technology. Our problem is that in many cases technology and its never-ending range of equipment possess people as they wade through reams and reams of data. We would be the first to admit the importance of good information in marketing or decision making situations. The quality of one's actions depends heavily upon good information served from data. We also believe that this information should be secured in the most efficient and timely ways possible.

Alas, the explosion of new technology worldwide.

The Return of the Body Snatchers serves as a warning to top managements of enlightened companies about the new generations that have only known very recent technology and the tools used to capture and display data. If indeed a company believes in a partnership with technology, a look inside the Pods would be useful. Our intuition tells us that the first part of the equation, gathering information, will be a strong skill set for young people. Our fear is that the second part—using the information effectively in the marketplace—will not be their strong suit.

A partnership with technology sounds easy, but in today's world it is in fact very difficult, given the proliferation of technology tools available. This partnership isn't just about computers or related data devices.

If you think the partnership with technology is easy, just turn off your cell phone, desk computer, laptop or BlackBerry for just a half day. See what happens.

A Bill Gates himself has said E-mail is a unique communication vehicle for a lot of reasons. "However, e-mail is not a substitute for direct interaction."

—from "Getting Wired: E-Mail From Bill"
(10/93-1/94) Quotationary, edited by Leonard Roy Frank (Random House Webster's)

A close examination of sales and profits in most corporations will consistently tell you that about 10 percent of your customer base produces a high percentage of annual profits. If you consider the sales force itself as a customer, about 10 to 15 percent of that sales force will produce 50 to 60 percent of your sales. The same holds true for charitable organizations or not-for-profit groups. And a few givers provide a majority of the money in political campaigns.

This basic tenet of sales life is consistent and tells us that customer contact can be prioritized, something we can put in the category of "very important to development and that probably can be accomplished." This can be done fairly easily in a reasonable time period.

Behind the duty to ensure the financial viability of the company, attention devoted to key customers by all key executives in the organization should be the highest priority. The annual growth and profit projections depend heavily on the performance of a relative few accounts and staff people.

Throughout the country in companies that have been excellent examples of "a high customer focus," a monumental breakdown is occurring, thanks to the Body Snatchers Syndrome. The large block of quality time that should be spent with these important customers has slipped dramatically, and the perceived value of critical customers is also sliding away in many organizations.

In most organizations, the need to develop additional information or data about key customers is truly nil, yet operations executives are finding more and more distractions and reasons to be inside—not outside. The focus has shifted to a circumspect one instead of one that looks externally. Justifications abound but the truth is that penalties will be paid for this behavior.

An example of a company doing it the "right" way can be found in Continental Airlines' penchant through marketing, communication, service and operations to dwell on the customer. CEO Gordon Bethune set a new mission for the company and that was to make safety, on-time service, and on-time baggage claim the tripod of sales strength for the company. Not long ago, after years of struggling to stay aloft, Delta's top management said that their airline had better "re-focus on our customer" . . . indicating that Delta's people needed to get a lot closer to their customers.

—*Ron Watt*

No data mining from data warehouses is needed for key customers. These tools are used for people you don't know. The irony is that more and more good customers have become strangers. It has become too easy for customer care people to stay in their Pods to study data, make plans, exercise their computers, and think about the bottom line instead of the top line. There is no good bottom line without a stout top line, and the way you get there is through customer care service.

Contact with key customers has become an occasional e-mail or voicemail or chance meetings at some industry convention. This treatment is necessary, of course, because these "customers are busy" and don't want to be bothered. In addition, "it is hard to get a hold of them or get them interested in our initiatives." Convincingly, the very opposite is true.

What a hell of a way to handle your most important assets!

Honda has earned a strong reputation in the marketplace for providing its customers with high quality products. Customer satisfaction is the "purpose" of Honda's Company Principle: "Maintaining a global viewpoint, we are dedicated to supplying products of the highest quality yet at a reasonable price for worldwide customer satisfaction."

Honda is a company that constantly tracks its customer satisfaction and quality indicators, using a number of data points. But, satisfying customers is more than gathering and reviewing statistics. Honda doesn't lose sight of the individual customer in the marketplace. Part of the reason came from Mr. Honda, himself. It was legendary within Honda that, in response to an exuberant engineering presentation proclaiming that a Honda plant had achieved 99 percent customer satisfaction, Mr. Honda roared back: "I am not interested in being pleased at 99 percent customer satisfaction! For the one percent who are not satisfied, we have produced a 100 percent no good product.

Here's an example of striving to achieve 100 satisfaction. In 1986, Honda of America Mfg., Inc.'s Marysville, Ohio, plant was undergoing a significant expansion—doubling its size. This included construction of a second paint production line. While important to meet growing demand for Honda autos, construction created dirt and dust challenges for the existing paint operations.

A customer in the Midwest complained to his dealer that the paint quality of his new Accord auto was not the same as his previous model. The dealer had tried to satisfy the customer by offering to have it repainted, but the customer was adamant that he wanted a car with factory paint quality. The customer wrote to Honda's Marysville's plant manager with his concern.

The plant manager immediately dispatched the manager of the paint department to the dealer to meet directly with this dissatisfied customer. As the manager tells the story: "I spoke with our customer, looked at our car, remembered our Company Principle and Mr. Honda—and on the spot told the customer we would deliver a new car to him that week. The dealer was amazed, the customer was satisfied and I knew I had acted appropriately to carry out our Company Principle. I also added the cost of one new car to our department's budget and resolved to re-double our efforts at achieving top paint quality and satisfying each and every customer no matter the challenges."

—Susan Insley, former senior vice president, Honda of America Mfg., Inc. Ms. Insley is currently executive vice president and principal, Cochran Group Inc.

"Would you like to hear some music while you hold?"

CHAPTER THIRTEEN

Service with a smile (a scenario).

RING . . . RING . . . RING

Thank you for calling the express service center for Body Snatchers, Inc. We are the leading manufacturer and distributor of laptops, desktops, on-line and off-line communication equipment and other devices to make your life both simple and efficient.

B. S. is recognized worldwide as the leading cloner of electronic products and services that you admire. We can give you the same wonderful feeling everyday that you have when you use B. S. products.

Your conversation may be recorded to insure accuracy of your request.

If you know your party's extension, press *8 followed by the four-digit extension number now.

Please listen to the following options for B. S. express service.

For a company directory, press 1 followed by the last four digits of the Poddie's last name that you desire to reach.

For questions relating to a billing you have received or other changes to be made to a billing notice, press 2.

For problems that you have encountered with one of our products you have purchased in the last 12 months, you can reach our help desk by pressing 3, followed by the last four digits of your social security number and the 15 digit B. S. product code.

Now press in your maternal grandmother's maiden name.

For warranty information, claims relating to product malfunction or the number for the B.S. customer care instant hotline, press 4.

If you desire an application for employment with B. S. Worldwide, press 5 and say send me an application and give your name and address in English, Spanish, French, German, or another language of your preference.

If you would like to purchase any B. S. electronic equipment including the brand new "Pod laser listening device" for $ 49.95, press 6.

To listen to excerpts from our CEO's recent annual meeting remarks titled "They make it, we clone it," press 7. The full text is also available, simply listen by pressing 10. It's really, really good. For ordering instructions press 11.

If you desire to make threats to our CEO, Mr. Watt, or members of his executive staff, first press 1 to secure full name and title and then press 8. A member of our computerized security staff will assist you. You should be aware that your conversation my be recorded for accuracy. Please note for significant threats we wish you to write to our corporate headquarters. You may obtain our address on our website . . . www.claimsan dthreats@bsproducts.com.

For a customer service representative to help with other service questions, press 0.

(key 1 is pressed by accident)

Hello, you have reached the desk of Skippy Jones, vice president in charge of B. S. Customer Service. I can not respond to your call right now because I will be out of the office for 10 days, until April 1st. Your call is important to me, so please leave your name and number and a brief message and I will return your call upon my return. For immediate service, press * 3, for the B. S. Help Desk.

Mr. Jones, this is Jeffery Stitz, a student at the University of Akron, in Ohio. I have been experiencing problems with your BosenBerry clone and cannot seem to get to the right person to solve a screen problem. I have tried to reach your help desk but the waits have been long and I have classes to attend. Please call me when you return.

Key 3 is pressed

You have reached the help desk for the entire state of the art B. S. product line. I am sorry that we have a heavy volume of calls currently, but your call is very important to us. Please stay on the line, and the next available Poddie will take your call. Currently, we estimate that your call will be answered in approximately 27 minutes, prior to our Central Standard Time noon hour. After that we must inform you that you may have to standby for 117 minutes accounting for our 90-minute close down for lunch. While waiting, we hope you enjoy the new CD just released by Rod Stewart, that will be playing in the background.

Key 4 is pressed . . . and sometime later:

Welcome to the B. S. Warranty and Claims Center. If your inquiry relates to one of our products that was purchased in the past 12 months and you wish to file a claim for replacement, **press *3** and one of our service representatives in the Help Center will gladly assist you with your claim. If the problem is with a product that is more than 12 months old, please find a claim form at www.claimsandthreats@bsproducts. com. Remember to include a self-addressed envelope to facilitate our response.

Key 0 is pressed

You have reached the Customer Care Department of Body Snatchers, Inc. Due to a high volume of calls at this time, we estimate that one of our service representatives will answer you call in approximately 39 minutes. While waiting, please enjoy the recently released CD from Rod Stewart that will be playing in the background . . .

It Had To Be You

Why do I do just as you say?
Why must I just give you your way?
Why do I sigh, why don't I try to forget?
It must have been something lovers call fate,
Kept me saying I had to wait
I saw them all, I just couldn't fall 'till we met . . .

It had to be you . . .*

Hello, this is Raja, how may I help you?

Raja, I am so happy to finally talk to a human voice! But you sound so far away; where are you located?

I'm right here in Chennai.

Chennai? I guess I don't know the city.

Chennai, India, but you can always reach me on Key 8. That is where I really reside in the B.S. system.

(losing signal, strange garbled sounds and a long whooooosh)

---

* From "It Had to be You," Gus Kahn-Isham Jones, copyright 1924

Key 8 is pressed inadvertently, just as the caller slumps over the phone, now quite asleep.

Back to oblivion.

CHAPTER FOURTEEN

Bad Signs, Good Signs.

FOLLOWING ARE 10 TELLTALE SIGNS that you have had your body snatched!

1. You can fix your own computer when it goes haywire.

2. You have not shopped for a new suit in five years.

3. Your cell phone has to be charged more than once a day.

4. You love your cubicle; it's like a nice bed.

5. You fret over attending black tie parties and charity balls; or you have never been to one or been invited.

6. You have stopped signing off your computer for days at a time

7. You feel your assistant is on his or her last legs. You can handle mail, messages, presentations and

appointments on your computer or BlackBerry. You don't have a clue as to how to work with an assistant.

8. You are feeling good with the way things are at work. Too much change is just not good! You are too comfortable—resisting change of any kind.

9. You have lunch in the company cafeteria with the same people every day.

10. You have never been inside a bank.

Now, you may want to take a look at this from a different perspective.

10 good signs that you have escaped!

1. Microsoft "Word" is a real challenge for me.

2. I think looking "business like" helps my image and the image of my company.

3. I keep a cell phone with me for "emergencies only."

4. "Working the crowd" is my calling and enjoyment.

5. I'm like a caged cat in my cubicle. I just need to get out there with customers!

6. I forget my computer password a lot.

7. My assistant is my joy . . . she or he is my computer expert, prepares my presentations, takes care of my "office work, and keeps my calendar going."

8. I like change; it creates exciting assignments and keeps everyone in the company on their toes.

9. I like meeting new people inside and outside the office every week.

10. I have coffee at the bank on Saturday mornings with the branch manager just to see what's new in the neighborhood.

Think about it. Who are you?

Don't worry. If you think you are in a Pod, we'll shortly tell you how to get out.

Real knowledge is to know the extent of one's own ignorance

—Confuscious, from Dr. Mardy Grothe's Oxymoronica (Harper Collins)

"Agreed?" "Agreed, 'Ol' Buddy!"

Oh my God! There are customers down there!

CHAPTER FIFTEEN

Watch out, there are customers all around you!

IT IS IMPORTANT THAT WE REVISIT the definition of "customers" as you re-unite yourself with their world. Most Poddies would tend to look at the customer definition narrowly, seeing them mainly as people who simply buy either goods or services from their firm. This is a natural thought process because the longer you spend trapped in a cubicle in Pod-life, the more you take on an "internal focus" as it relates to a customer and customer dynamics.

Now that you are escaping the Pod and have taken on more of an external focus to things, including customers, the definition becomes very broad. Suddenly the customer is anyone who needs to be "touched" to accomplish both narrow and broad missions.

Most certainly, to become more externally focused you ironically have to relate more effectively with your fellow "inside" employees—especially those critical to your mission. These insiders are customers and influencers as well. Try, try, try to make these once infrequent contacts "face to face" instead of using e-mail or voicemail. Insist on seeing them in person, and please do not fall back into the trap of posing or answering questions through e-mail or voicemail. Non face-to-face contact misses the point at hand completely. If it can't be face to face, at least make it a "live" phone call. If it involves a group of people, and video conferencing is available, try that. At least this is a "live" interactive device. Without face-to- face connection or a suitable other form of spontaneous, live interface, the point at hand is often missed completely.

We do *not* disdain e-mail or voicemail used wisely. A good e-mail or voicemail ploy is something that will pique the interest of the receiver. "I have a couple of new ideas that I would like to discuss with you," or "I hope you can take time to help me develop a new service technique," or "I saw a recent article on branding that seemed to be a good idea for our company." "Can we get together soon to discuss?" All of these techniques seem to work to get some "face time" with key internal customers. The secret is to not get into an e-mail or voicemail dialogue that eventually gets you shut out of a face to face visit.

As you recuperate yourself from the Pod life, work naturally will have more personal contacts with fellow employees. You need to be the modern "Dr. Miles Bennell," who warns fellow employees about the dangers of becoming possessed with life as created from seedpods . . . no heart, no soul, no emotion. And *safe and secure and weak, complacent, tedious, predictable, boring, loutish, anti-social, changeless and risk averse.* The more personal contact, the more confidence you build within yourself, and it stimulates others to have more confidence in you. Interrupt their dull day to congratulate them on a son or daughter's accomplishment or simply ask their advice on a pending project or a problem to be solved or an opportunity to be accomplished.

A good marketing tip when you initiate these contacts is to always talk about them instead of you. This basic marketing "rule" applies to customers at every level both inside your organization and outside.

It is difficult for fellow employees to "get to know you" through e-mail or voicemail, the all to frequent everyday communication tool. That "getting to know you" is critical if you have thoughts of advancement in your organization. If you are a young trainee or a budding executive, you need to find ways to see executive management "face to face" or at least to make sure your face is recognized in the company by a broad spectrum of officers and managers. Find ways to expose your talents including outside activities, such as volunteerism, working for political candidates, lobbying for institutional projects that you feel strongly about and more. Also, do not be shy about making positive suggestions on current company operations and ask if you can be a part of leadership and other continuing education programs.

When I was a part of the executive management team, I had the opportunity to spend a considerable amount of time with young executive trainees and had many chances to interact with them, many of the contacts initiated by me. These relative new and somewhat naive employees were always concerned about "getting lost" in a large organization, like ours that had talented people who were very competitive.

I always found it to be true and told them frequently that it is very difficult to lose or not recognize outstanding employees, young and older, in a small, medium or large company. This is even truer if the executive team running the company is expert in developing capable succession throughout the company. The fact is that outstanding employees in most companies stand out like a "sore thumb" despite every attempt by supervisors to hide them!

An important part of internal recognition comes from extra-curricular activities, not just what you are accomplishing in the workplace.

As some of our young employees decided to breakout from their cubicles and to become more externally focused, the more often I had the opportunity to hear them give presentations inside and watch them take on key community activities outside the organization. I watched them become leaders!

Soon, you would see their names appear in several key "boxes" in the future organization chart of the company.

—Cary Blair

The "second" group of customers that demand your attention is the distribution or sales team for your products or services whether they are captive or independent. This is the heart and soul of company success, top line and bottom line. This is the hardest job in the marketing chain, yet the most rewarding, if you have a passion for the face-to-face contact. The sales people should be the idols of employees in an "externally focused" company. *They are capable of making good things happen with something ordinary and, conversely, they can make bad things happen with something extra-ordinary,* if communication goes lacking. They are often quite amazing people, and they should be revered for their skills and talents. If you communicate with them consistently, good things are bound to happen.

For a Poddie who has escaped, it is essential that you spend quality time with this group of "customers." There is little to learn sending e-mails or voicemails to this group! They want to look you in the eye, they want to feel your pulse and they dare you to say "no". They are your "trusted advisor" and they want to reinforce that feeling with you personally, at every opportunity. This is the epitome of external focus and this will become the driving force about a desire to avoid cubicle life forever.

I spent many, many waking hours with this group of customers, first as a marketing rep early in my career and as CEO later, and nothing has changed. We enjoyed phenomenal support from our independent sales group, our agents, throughout my career. I believe this was a direct result of our willingness and desire to spend large chunks of time supporting them "personally" at every level within the company.

The feeling also always existed that we could go downhill quickly if that personal face-to- face contact started to erode. I have always felt that an internally focused organization spent too much time thinking about "me" or "what's good for us." Externally focused organizations think an awful lot about "what's good for you, Mr., Ms. and Mrs. Customer." I liked the "you" part and practiced that customer focus throughout my career, and still do now in my other endeavors.

—Cary Blair

Finally, let's not forget about another large customer segment that is often overlooked in the rush to make sales or cut expenses. We break this customer set into two groups. The first is the increasing numbers of consultants and vendors that have become a big part of every company's existence. It is practically impossible today to do all internal development of systems or to avoid regulatory compliance or independent auditors. Each of these needs brings new non-employees to the premises.

Often they are treated like what they are—strangers that come and go.

A good marketing organization will treat these people as "insiders" and will induce these innocent bystanders for hire to help "make the plan." They not only hear the mission but also are asked to get intimate with short and long range goals of the company. Project leaders discuss tactics with them and ask for their advice and counsel based on what they have seen in other organizations. Attitudes change, cooperation gets more intense and a win-win posture occurs on both sides.

Marketing is about creating disciples for your mission and what's wrong with enlisting outsiders that are on premises to help develop new systems or to keep you compliant? Is there a chance that they might say something positive to business friends and acquaintances that leads to more business for you?

The other often overlooked set of customers is state, local and national political officials, not to forget community leaders in places where employees are hired or where the company does business. Included in this group also must be the media, who have an increasing role in the way the image of your company is portrayed to the buying public and the rest of the outside world. In an organization that features an internally focused management, there is complete disdain for much of this segment of customers. Some might describe them as a "necessary evil."

A good marketing focus can turn this group of customers into some of your strongest supporters and many times it does not depend on how much money you have donated to their campaigns. We have said this many times, but success with this group also depends on your willingness to focus on "them," a cardinal rule for good marketing successes. The basis question is what is important to "you, Ms. or Mr. Official" and how might we help accomplish some of "your" goals.

With a good "customer focus," your company will be lauded for its volunteer spirit to communities, being a good corporate citizen or aiding in the success of important local, state or national initiatives with fundraising campaigns or other kinds of help and support. None of this recognition has to come from internal advertising campaigns; it all comes from your new "customers" talking about you to other potential customers. What a way to turn a "necessary evil" into one of your best sales persons!

Customers all around me and you!

"*Of course you're going to be depressed if you
keep comparing yourself with successful people.*"

CHAPTER SIXTEEN

So you want to stop being a Poddie?

HERE ARE SOME COMPETENCY THOUGHTS for potential Poddie converts, and key attributes that we look for regarding management succession in various parts and levels of the corporation. These are the human assets that have much to do with the future success of the organization as well as your personal success:

- Steps up and can make tough decisions . . . pulls the trigger!

- Good communicator, good listener and likes the interaction with people.

- Offers innovative ideas and effective solutions to critical operational problems.

- Understands his/her personal weaknesses and builds a complementary team to fulfill the tasks at hand, covering individual weaknesses.

- Knows how to build trust among the people around him/her . . . very creditable.

- "Vision" seems to be his/her forte . . . their thought process just seems right for the organization and its future success.

- Has good timing. Seems to know what is right for the organization at the right time . . . asks penetrating questions and exercises good judgement.

- Has the "Arnold Palmer" factor, the magnetism. Exercises "people skills" . . . people just want to be around this person as a mentor and as a leader.

- Has a deep understanding of how the company makes money and appreciates the contribution of each individual department to that end.

- Demonstrates "marketing skills" to the extent that she or he helps create a "customer focus" in the organization.

- "High energy" person who is results oriented. And knows how "to keep many balls in the air" but has an excellent sense of priority.

- Has an openness to new ideas and is willing to change his/her point of view.

- Smart person in terms of I.Q. but, more so, has "street smarts" . . . intuitively perceives patterns of external change and can adapt "vision" to these changes.

- Has "quick study" characteristics. Is always curious, always an observer, who acts rather than reacts.

- Has a track record of personal success. Strong record of extra-curricular activities in college and in career, helping to make her/him a well-rounded individual.

- Understands the concept of "value creation" and the relationship to the top line and bottom line.

- Has high moral/ethical standards and brings a sense of integrity to the company.

- Has a track record of removing "barriers to change" and is not content with the "status quo." In most cases believes that "sacred cows make the best hamburger."

- Loves interaction with "stakeholders." Loves the job, loves the people, loves spreading the word.

- Has a "life," not just in the work place. Is well read, knows how to educate herself/himself well beyond formal education. Is up on current events. Has a handle on what's happening out there.

- Knows how to balance work life and personal life.

"Few things are harder to put up with than a good example."

—Mark Twain, from "The 2,548 Best Things Anybody Ever Said," compiled by Robert Byrne (A Fireside Book—Simon & Schuster)

"With you, it's different. You've got talent, courage, imagination, savoir-faire . . . "

CHAPTER SEVENTEEN

"Guerilla" Marketing 101.

FARIBORZ GHADAR, "F.G." TO HIS FRIENDS in the business world, directs the Center for Global Business Studies in the Smeal College of Business at Penn State University. He is an engaging speaker who seems to always make his point about global competition and marketing focus, using real experiences, things that happen personally to him in many instances.

One of our favorites is his experience with LL Bean, the hands-on marketer of sporting goods and sportswear.

One day he called Bean to buy his sister a birthday present, a new pair of waders to use in her favorite trout stream. What ensued is a good marketing lesson for all of us, and it shows what smart customer service is all about:

"How you treat your customers can be *the* determining factor in how profitable your business becomes. High customer satisfaction levels are critical to the overall success of your business, regardless of the overall value of the product or service you offer. In other words, your customers have to love *you* as a corporate identity.

"While technology can assist you in creative ways in providing your product or service to the marketplace, it is your employees who are really the soul of your organization and can be the difference in your overall success. However, in order to perform this pivotal role, they need to have the guidelines clearly articulated for them and feel comfortable enough and empowered to express your organization's loyalty and dedication to its customers. I had an experience with LL Bean, when I first moved from Washington, D.C., to State College, Pa. that proves the significance of achieving this symbiosis quite clearly.

"My sister Margaret's birthday was fast approaching and I asked my wife Lis for some ideas for what we might get her as a present. Now my sister has very particular taste and she already has everything that she could probably ever want, so buying gifts for her is not that easy. After thinking about it, Lis remembered that Margaret had mentioned at one time how her fly fishing waders were getting old, so replacing them seemed like a perfect birthday present.

"When Lis left to go to the supermarket, I decided that instead of unpacking more of our boxes that I would go ahead and order Margaret's present. I naturally thought of LL Bean, but didn't have ready access to one of their catalogues, so I called information for their number. When the LL Bean representative picked up and introduced himself, he asked me for my name. When I told him, 'Ghadar, Fariborz Ghadar,' he immediately replied, 'Mr. Ghadar, are you traveling today?' Since we had just moved, the telephone number that had shown up on his Caller ID screen must not have matched their records with my account. He then asked me to update my contact information for him. After this was quickly done, I asked him to go over what kind of fly fishing waders they carried. He proceeded to tell me about their models and I picked a size and placed the order.

"I went to my new office, and forgot to tell Lis that I had placed the order when she came back from grocery shopping. Being the efficient person that she is, she found the box with the LL Bean catalogue and proceeded to call the company to take care of ordering my sister's birthday present."

LL Bean (woman's voice, typing in the background): Hi Lis, this is Sharon. So how was your move to State College been so far?

Lis (surprised that Sharon knew about our move, replies): Hi Sharon. We are still living out of boxes, but the area is so beautiful and we are just getting to know some of the things to do and see around here.

LL Bean (still typing): Did you know that only six miles from your new house is an excellent fishing spot? Former President Carter used to go fishing there all the time when he could take time away from the White House.

Lis (laughing): Really? I will have to check it out when I have finished moving in.

LL Bean: You definitely should. But you know that the fishing equipment you already have isn't appropriate for this type of spot? I can certainly recommend the type most appropriate for this location.

Lis: Thanks, but not today. I have enough to do right now without trying to fit in a day of fishing. What I really was calling for was to order some new waders for my sister-in-law as a birthday present.

LL Bean: I see that an order for fly fishing waders was just placed by your husband only 20 minutes ago. I just want to verify that this isn't a duplicate order.

Lis: Well, it must be a duplicate, so go ahead and process his order . . . although I can't believe he was so proactive!

LL Bean: While I bring up the order screen, let me tell you about some of our new specials. (She proceeds to go over their promotional items and of course Lis buys a few of the *special* products.)

Lis: Since I have you on the phone though, I might as well order a pair of rubber boots for my husband now that we have moved North.

LL Bean: (realizing this is a duplicate order from last Christmas) Does he like the boots you got him for Christmas? Is he happy with them?

Lis: Actually, he loved them, but after only three months, they had a tear in them.

LL Bean: Oh, that isn't good. Let me go ahead and send out a new pair right away to your new home in State College, because he will be needing them with the weather in that area. This is, of course, complimentary, as this shouldn't have happened at all. Sorry about that.

Lis: Thanks so much. That would be wonderful. By the way, what was the name of the fishing spot Carter would go to again? I just want to write it down so that I can ask some folks around here about it.

(They continue to discuss fishing and what kind of gear and special hooks Lis would need for the State College area, and Lis proceeds to place an order for all new fishing equipment so that she will be able to go to this acclaimed fishing spot.)

"Now this experience shows some very important things about LL Bean:

"1. Lis didn't complain about the boots being torn, but was offered a replacement pair anyway, for free. This is LL Bean's policy.

"2. While all customers can return products that did not work for them, only their high-value customers are tracked so that even potential problems not articulated are addressed. Clearly, I am not a high-value customer as I rarely order anything and if I do, it is usually under $50 including shipping. Lis, however, orders a wide variety of items, and if given any incentive, will order more than she initially was intending to when she calls. So she is not only tracked but also offered numerous specials on the phone interactions.

"3. When I called in, the customer service representative didn't pretend to know where I was calling from, even though he could see from the area code where it was. Instead, he asked me if I was traveling. With Lis, the customer service representative initiated friendly conversation right away, and then proceeded to chat for five minutes. They obviously have a privacy indicator that shows that I am paranoid about people knowing too much about me after having fled the Iranian Revolution, whereas Lis was born outside of Philadelphia and shares her life story with everyone. It is critical that firms have a clear idea of what the customer's privacy preferences might be.

Different people prefer things in different ways, and companies need to always be aware of that.

In France, for example, consumers want to feel like their personal information is kept private, even if it just means that the customer service representative pretends not to know the information about the customer. However, in Mexico, the consumers' attitudes are completely the reverse. AT&T's local partner had a much loved campaign there where when a customer's eldest son's birthday came around, AT&T would call and give the family their congratulations on his birthday and offer them a gift certificate for a free five minute telephone call anywhere that day. The

Mexican consumers loved this personal involvement and attention.

"4. Lis was offered additional promotional items she didn't already have, but since LL Bean knew that I don't purchase in this way, they didn't bother to tell me about any of their specials. LL Bean also knows that they already have a high percentage of Lis's wallet share. When they offered her an item, if Lis had responded that she already had the fishing gear suitable for the new fishing spot and she said that she had acquired the gear from somebody else, LL Bean would have inquired if she were happy with the product. Then, Lis would have been offered a competing or complimentary coupon in order to win back Lis's loyalty.

"5. The LL Bean customer service representative felt empowered to give Lis a free pair of boots. Even though Lis didn't even ask for them, or complain when the topic came up. Their high-value customers are treated with even more exceptional service.

"6. The net result is that the customer feels very comfortable buying from their company. LL Bean adjusts accordingly to the type of customer who calls in—whether it is someone like me who only wants to get prompt and efficient service, or someone like my wife who enjoys the whole process of getting to know the customer service representative. LL Bean has created a corporate culture that ensures the customer feels personally taken care of.

"Two years after this parallel but satisfying experience that Lis and I had with LL Bean, we were on vacation in New England. We were more than 150 miles away from LL Bean's headquarters in Freeport, Maine, but Lis insisted that we drive out of our way and see their operation (and do more shopping!).

"Lis dropped me off at the front door and went to park the car, while I went into their retail shop to put my name on the list for the general tour and wait for her.

"But when she got to the front door, she went to the VIP area for their special *in-depth* trip. As soon as she asked about me, they came to get me out of the general public line and added my name to Lis's on the high-value customer tour list. Without her, I would not have individually qualified for such special treatment.

"Even though LL Bean did not create the technical systems that they use to interface with their clients. They know how to pinpoint a customer, using these sophisticated systems to the mutual benefit of the customer and the company. Clearly, though, it is not the systems alone that can accomplish such high customer satisfaction and loyalty; more so, it is smart people operating within a corporate culture that *wants* to satisfy its customers. That is what makes LL Bean so successful and profitable year after year.

"Satisfy the customer and profits will naturally come!"

*We're expecting you to make a birdie or eagle on
our sales charts this year, Jim.*

CHAPTER EIGHTEEN

Getting results from doing the unusual.

ONCE UPON A TIME there was a very old and wise insurance company that sold their products throughout Mid-America. They were much understated in everything they did, from advertising to charitable giving. They employed an independent sales force of agents that would mention the company's name from time to time in the selling process and cite its strong credentials to the potential customer. Once on board, the customer or agent seldom would leave.

Despite its size, shape and branding timidity, the company was a marketing giant, with hands-on touch and feel. It had a do-it-right attitude with an unrelenting customer focus. Employees who grew up in the company knew that customer care came with the territory and spending considerable time with customers was number one in its job description. This philosophy was imbedded in the fabric of the company and its many people and agents.

The company also was a charitable firm and a good corporate citizen that gave back often to local, state and national causes. Since the company was founded in 1848 and was headquartered in a rural area near Amish Country, its philanthropy came naturally over the years.

Employees could be seen building houses for Habitat for Humanity or providing assistance to residents in low-income areas, or volunteering in a multitude of other ways that would help the surrounding community. In fact, the Governor recognized the employees of the company as the top volunteering corporate group in the State in terms of numbers of people working and the hours of help they provided in community projects.

While speak of the company's past the goodness here is that its tradition of customer and community focus and volunteerism continue today and have set the stage for another success—a recent one.

In the early 1990s, the company embarked upon a more upscale campaign that would see it further recognized for the strong value system that had held the organization together for more than 150 years. An awareness effort was launched that was focused on young men and women, the next generation of leaders—and, of course, customers.

The Professional Golfers Association of America (PGA) had for years been courting young people to enter the game of golf through various annual promotions and through local and national tournaments. After all, golf was a growing game in the 90s and these young people represented the livelihood of PGA professionals into the future. For 25 years, the PGA had conducted a national golf championship for juniors at various venues utilizing various sponsors.

In 1990, the tournament was begging for a sponsor— at a time when junior golf competition was exploding across the nation. In fact, teenagers were being invited or were qualifying to play in national amateur and professional events. At the time, the PGA Junior Championship touched many young people but was not being recognized as a leader in competitive golf, or as a teaching tool, by the players or their parents.

The insurance company owned private golf facilities that had been used since 1936 for considerable customer entertainment and employee recreation. The facilities ensured a "personal touch" that was much like entertaining in your home. Company officials made a call on the PGA of America to determine its interest in an "unknown" becoming the sponsor of the PGA's national championship.

The company said it wanted to set into motion a program to teach young people about life experiences, competition, benevolence and ethical values that could be used the rest of their lives. Company officials made it clear to the PGA that if it moved the championship to the company's facilities the tournament would easily be recognized as one of the "junior majors."

Being the best was part of what the company's tried to do every day. After a five-year contract was signed, this quiet, hands-on customer care company set the new junior golf marketing plan in motion.

Although the company had never conducted a major sports event, it had entertained customers in large groups for years and this golf event was clearly all about customer care. Volunteer working committees were formed for every aspect of the process. It would host 41 national qualifiers and a national championship in partnership with the PGA each year.

Soon the volunteer list of employees, club members and customers reached over 300. A solid marketing plan was starting to take hold. Volunteers and employees had bought in to the excellence theme that trailed the company in everything that it did and this was rubbing off on the new junior golf "customers" it was now touching.

Spectators held the key to the second phase of the marketing plan. This was junior golf, ages 12-17, not the biggest draw in the world, but some household names were being created in junior golf, despite their young age. Observers from the insurance company noted that the only spectators at major junior golf events were mom and dad and sometimes a brother or sister. Grandmother and grandfather would rarely walk the course to cheer for good shots.

Lots and lots of spectators were needed, the marketing plan said as a lure for more quality national sponsors, media coverage and the college coaching community that wants to see what the competitors will do under tournament pressure. More than just mom and dad had to be at the event. The company wondered how it could turn 100 spectators into thousands.

The company and the PGA launched a media awareness campaign that highlighted America's young new role models and the values that can be learned through competitive golf. The company and the PGA emphasized that golf is the only competitive sport where the players call rule violations on themselves. Ethical conduct was certainly a valuable life skill, given some of the abuses that have taken place in U.S. business.

Company officials played golf, hosted lunches and dinners with the media officials, writers and broadcasters and prepared easy-to-use information packets as part of the execution of the marketing plan. A public relations consultant was added to further enhance marketing and communication.

Soon, as the number of media interviews mounted, the name recognition of the company and sponsors heightened. Ticket demand increased as well, as did joint sponsorship presentations. Moreover, nearly 6000 spectators attended the 1991 four-day championship. Kids came in planes, busses and vans with their parents and, for the first time, their grandmothers and grandfathers. Fourteen-year-old players could be seen signing autographs for eight-year-olds. Players, parents, grandparents, spectators and media were in awe of the preparation and execution of the first year's event at the company's golfing facilities. Volunteers and employees had vowed to make this a lifetime experience for these young people. It seems the company's new "customer care" tactics were working.

Next came the hands-on part. "Customers" love this part, at any age! Everywhere the contestant went, he or she saw their name . . . locker room, practice range, scoring placards, pairing sheets and on the massive scoreboard on the 18th hole. Every evening a small gift was put on the player's pillow as a thank you for coming and helping to "make the week" for spectators, volunteers and staff that had worked so hard.

Parents, who sacrifice time, money and energy for these talented kids, were treated the same way. They received daily meals in the company's inn, courtesy cars from a sponsor, and housing in the area at discount rates.

Playing conditions, services and facilities were extra-ordinary—the entire company organization realizing they all were part of the "marketing staff and grounds crew." They kept everything on a championship level.

The CEO of the company circulated every day, talking to players and parents to ensure that they were having the experience of their lives.

The scope of the endeavor saw people coming in from 50 states. Regional qualifiers for the championship were held throughout the country for boys and girls. Interest in the 120 championship positions began to grow . . . as many as 9000 young people were trying to qualify. The company sent some of their star employees to all 41 qualifiers to emphasize its genuine interest in these "customers." They assisted PGA officials in many aspects of the tournament even though many had never played golf. Being a good marketer does not always depend on deep knowledge of the product. The effort became a valuable training exercise for employees in the planning and execution of a creative marketing campaign.

The expense of these annual qualifiers and national championship was no small matter, therefore a campaign to attract additional sponsors was critical. Company executives began to make personal calls on regional and national companies that were highly respected and had a mission to help develop young people as qualified leaders in the future. Competitive golf was the attraction but setting examples that kids and their parents could "see and feel" was the long-range goal. Soon the sponsor board began to fill up. They could also see the "vision." Following tournament week, sponsors started to get letters and accolades from players, parents and spectators quite unlike anything they had been involved with previously. Letters always mentioned their products and their company and offered many thanks for their participation and belief in today's young people.

Ho, ho, ho, back to the CEO. Every day of national tournament week, out of the north came this old white-haired man, riding his Club Car, ladened with gifts of tees, socks, sunscreen and golf gloves. He worked the crowd relentlessly, the old man's specialty. He wrote down names and numbers and he consoled players that had less than a stellar round the day before. He encouraged everyone to have fun. He congratulated those who had played well. He gave interviews and thanked sponsor after sponsor, media person after media person.

Players and parents were merry about the little gifts the players found on their pillows and the meals enjoyed every evening. Suggestions were solicited for improvements in "their" tournament. Months later, players and parents would get another small memento and a personal note from the old man thanking them for being such a good role model for their peers and inquiring of the kids about their academics or their college plans.

Today, more than 400 former players stay in touch with the old guy, because there might just be one more lesson to be learned about creating strong values in one's life and thinking about excellence in everything one does.

In junior golf, these young competitors and their parents will forever know the white-haired old man and now former CEO of the company as "Cary Claus."

Being the customer's "Santa" isn't all-bad! End of the marketing tale.

"You smell like a chimney."

Let's go in for batting practice and a beer.

CHAPTER NINETEEN

Reverse motivation is not a good idea: the real meaning of team.

In nearly 40 years in the public relations and marketing field I have witnessed many situations where people in management did it right—and did it woefully wrong as well. These vignettes should make some points about hands-on, face-to-face effectiveness or the lack thereof.

—Ron Watt

Billy Martin, the late, often great manager of so many major league baseball teams, such as New York, Texas, Detroit, Oakland and Minnesota, had this uncanny ability to take a ragamuffin club and turn it into a gem. Later he'd get bored or unpleasant and would find himself fired. But the one thing that sports buffs will always remember is that this little Italian guy from San Francisco could put something together that just about no one else could. If anything, he was one of the great face-to face operatives you could ever meet.

He might be at Toots Shor's in New York or The Theatrical Grill in Cleveland and always, in addition to the complement of drinks proffered by the customers, he had a few young ball players nearby. He'd be giving them batting tips and stealing tips and fielding tips. They'd listen attentively. Some of them, people like Chris Chambliss and Graig Nettles, became some of the game's best players of their day. One might say Billy was always working, but he was always having fun as well. People listened to him, because he had such an interesting take on the game and life.

Billy was just a wee guy who was the second baseman for the great New York Yankees clubs of the 1950s. His teammates were the likes of Mickey Mantle, Yogi Berra and Whitey Ford, Hall-of-Famers. His manager was also a Hall-of-Famer, the avuncularly redoubtable Casey Stengel, who was a great face-to-face man himself. Casey was once so upset about one of his players having an inability to slide properly, Case took off through the grand lobby of the Waldorf Astoria and did a hook slide into a sofa containing some blue-haired matrons. The player got the message. Casey was well into his sixties when he demonstrated that proper sliding technique.

These are the kinds of people who stand out, because they would never practice their trade from within convention. In the day of gadgets and devices they would still do it the same way, because they would still have an affinity for face-to-face contact to make their message clear. To make their selling proposition.

On the industrial side of things, one of the most remarkable men to be called a CEO was Fred

Crawford, who took the old Thompson Products Co., which made automobile and aircraft parts, and turned it into TRW, the Goliath aerospace company. He also had such a loyal employee base that union organizers could never have their way.

Once he recollected, "A guy who looked like a football player, a big guy with no neck called on me from the Steelworkers Union, and he told me he wanted to have me sign a contract for a 'closed shop' at a new plant we were building in Pennsylvania. He said the union had a lock on every plant that operated in the State. I asked him what would happen if I didn't sign the contract, and he said 'skulls would be broken.' I thought here we were in the depths of the Second World War, thousands of people losing their lives to protect our liberty, and this mug is telling me we had to have a closed union shop. I asked him why couldn't the workers determine what they wanted, and he said they had no choice. I showed him the door, and the new plant was never organized. My theory was that sometimes attack is the best defense. We convinced the workers in Harrisburg that they would be better off without the union contract—and they were."

Fred Crawford was renowned for knowing his people. The company was their company as well as his. When he walked the floors of the company's plants, as often he did, he knew each employee by name and knew about their families. He talked with them personally and as he did he'd weave in pertinent company information he wanted to convey. He was hands-on, and what he did can still work today—but really not that well from your BlackBerry.

Sometimes you have to wonder what the hell ever happened to common sense.

I once worked with a CEO of *a large company who each year would send in his company's dues of $50,000 to the regional economic development organization. He said, "That's it. I'm not going to any meetings and I'm not wasting my time with this organization; I'll just take care of our financial obligation."*

We had been advising the executive that it would be of benefit to his company and his 2,500 local employees to do good spadework that would be beneficial to the community. Create some noticeable and meaningful projects that would shed some good light on the company and him. Every city can use the might of its corporate citizens in education, economic development, race relations and government/industrial cooperation, to name a few things. The guy just couldn't be persuaded to see the light. His stance was: "Look, I don't know why or how busy CEOs would take the time away from their jobs or their families to do these things. It's stupid."

Eventually, the company and this guy disappeared. He and it might have lasted a lot longer if he had taken a more rounded, holistic approach to business. Instead, his ennui inspired little loyalty from the employees and community and, essentially, he just did not perform the many tasks of a CEO effectively.

—Ron Watt

Max DePree wrote a book a few years ago likening a company that operated fluidly to that of a big jazz orchestra. If you prefer the classics, the same can be said for a 100-piece symphonic group to a 20-piece jazz band. The drill is that everybody has to be on the same page. Sure, you want solos from the appropriate virtuosi but in the end the whole organization has to finish the piece together. This takes the inspiration, talent and sometimes genius of the CEO or the conductor.

One of the best CEOs I ever met wouldn't be thought much as one, but of course he was. His name was William Basie from Red Bank, New Jersey. Most people knew him by his nickname,"Count." Basie found he had to go West to get on the map, so he went to Kansas City, where he eventually founded his own band that swung like no other before or since.

Count would sit at the piano, seemingly not even playing, but he was. You'd surely notice that if he walked off the stage in the middle of the number. But he just had the immense common sense to let his men play; and he held everything together, without flourish or brazenness, more quiet than grand. He had a band that kept its players for a veritable lifetime. There were many talented leaders in the music, but no one had the serenity and wisdom of Count Basie. Yet isn't it strange that what came out of his orchestra swung, kicked, bayed and wailed in its unmistakable style, truly showing that the whole was larger than the parts? His arrangements are still driving players and listeners wild to this day; they are timeless. Sinatra, the other Chairman of the Board, thought he never sounded better than when he sang against the backdrop of the Basie band. You had to be damned good to be in the

Basie band, and if you could get in it you would never want to leave.

But the Count would modestly say, "You think I'm the leader of the band . . . no, no, not me. That would be my young friend, Butch Miles." In the 1970s, Butch was the twenty-something prodigy drummer of the band, the average age of which was close to sixty. All the players but one were black. One was white, Butch Miles, and he sat at the very helm of the rhythm section in the swingingest band in the land. Count Basie always, always thought out of the box and he never needed a Palm or BlackBerry to create his work and run his musical organization.

By the way, Count Basie has long passed on. But guess what? Butch Miles is still playing the drums for his magnificent orchestra.

—Ron Watt

A long time ago, a company that was on the road to great success, Owens-Corning Fiberglas Corporation along with two of its public relations firms, Burson-Marstellar and Flournoy & Gibbs, started a program called "Youth Drop-In Clinics." It turned out that this venture was so successful that it won every significant national award in the manufacturing and public relations industries. It was beamed at elementary and secondary school students during the summer months, when they had too much time on their hands.

The way it worked was like this: clinics, featuring star athletes of every sport from baseball, football and basketball to track, tennis and auto racing, were held in the mornings and afternoons in inner city and suburban areas throughout the country where OCF had a stake. At lunchtime, the athletes were further utilized as headline luncheon speakers for the company's customers, distributors and sales people. The kids and the adults both loved to meet and hear the likes of Oscar Robertson, Paul Warfield, Big Daddy Lipscomb, Andy Robustelli, Mario Andretti, Clark Graebner, Billie Jean King, Rosie Casals, Evonne Goolagong, Billy Nelsen and many others.

At the luncheons, the sports stars talked about not only their careers and teams but also key Owens-Corning products that the company was pushing to its marketplace. Everyone benefited.

For the kids, clinics were run about a particular sport and the kids got to participate as well. But in addition to this, each clinic emphasized the importance of staying in school, being a good citizen and learning how to play the game fairly and effectively.

Like Westfield Insurance and Westfield Financial Corporation, which have so effectively hosted the Junior PGA and Junior Ryder Cup championships, Owens-Corning showed it cared about young people —who also, by the way, would become its future customers. Face-to-face, hands-on contact at its best and that which had many dividends.

Another company that has been impressive over many years is RPM, a name you might not know, but you certainly know its many products in the consumer and commercial marketplace. Products such as Zinsser, Testor's, Bondex and Day-Glo. Over the years, RPM acquired a number of family-owned companies, and unlike it so often goes, their policy was to *keep* the former owners and key executives in place and provide them with strong incentives to keep their units growing and successful. RPM has long been a favorite of stock clubs and institutional investors alike. It is efficient and nimble, with a small corporate staff, and a hands-on approach throughout the organization.

It too started as a small family operation, called Republic Powdered Metals. Today it has $2 billion in sales, owns units throughout the world, and is traded on the New York Stock Exchange. It is now in its third generation of the Sullivan family running its operations, and with each generation, the company has reinvented itself and girded itself for continued creative business success.

At our own firm, Watt, Roop & Co., in the middle of its more than twenty-year run before becoming part of one of the world's largest PR firms, Fleishman-Hillard, we tried something that almost proved to be a disaster.

Someone had a bright idea that seemed a bit dubious to me, but I went along with it, thinking this might help us with growth. I should have trusted my better judgement.

We thought we'd break the company of some forty people into three client groups. Each of these groups were responsible for about 15 clients. Prior to that, the company operated pretty much as one good-sized team, all pitching in for the overall cause. Now we had three groups literally competing against one another for clients and financial gains. We tried it for about a year, and then decided this was good way to wreck a good company.

Human nature being what it is, the three groups cared more for what was going on within their groups than with the macro picture of the company. People who were once teammates and friends pitted themselves against one another. Essentially what we created was three agencies within the agency—"all for my group, all for me" became de rigueur. I've never seen so much tension created in a fairly small company over such a short period of time. "This is ours!" became the mantra. Petty jealousies and envy seemingly occurred around every bend. Client service and new business production got mired in confusion. We were like a runaway conastoga wagon. No one benefited.

We scrapped the idea before it scrapped us, and we re-created what a small to middle-size consulting firm out to be—a team!! We once again spurted in growth and went on to become one of the largest PR counseling firms in the Midwest. Some ideas just don't work, and the smart leadership will know when to cut their losses before they are dealing with a shambles they can't rectify.

—Ron Watt

Just remember whatever the department in which you reside, there is a bigger picture to always be aware of. You should be aware of the total team's goals, obstacles and opportunities. Wherever you are in the company you have to know the company won't last very long if everyone, everyone is not customer friendly. Personal and departmental ambitions should never override the basic mission of the organization. When the company succeeds, you will have the best opportunity to succeed.

Being buried in your personal contraptions and the desultory art of avoidance gets you nowhere in the company or outside. Get out of that Pod and meet everyone you can—the in-house "customers" as well as the customers and clients the company relies upon for its business. They are by far the most important assets any company has and they should know they are—from everyone with whom they come in contact up and down and sideways throughout the organization. Looked upon this way, there is not a person in the company who is not a salesman or saleswoman.

Do it face-to-face, hands on, not from behind some inert, dehumanizing screen.

"You'll love it! It's computer-driven but people-enhanced."

CHAPTER TWENTY

Spend lots of quality time with key customers. These <u>percentages of success you should know about.</u>

WHO ARE YOUR MOST KEY CUSTOMERS? It's a good idea to find out. These are the people with whom you should spend the most time, because they will be the most effective for you in building your success out of the Pod.

Remember that 10 percent of customers produce the highest percent of annual revenues and profits. And never forget that 10 to 15 percent of your sales force will produce 50 to 60 percent of your sales. Keep in mind that this maxim holds true for charitable organizations or not-for-profit organizations, as well as for the for-profits. A smallish percentage produces the most!!

This basic tenet of sales life is consistent and tells us that customer contact can be prioritized on a scale of very important to developmental and can be accomplished fairly easily with some effort over a reasonable time period. Beyond the duty to ensure financial viability of the company, attention devoted to key customers by all executives in a company—yes, by even the financial people—should be the highest priority. The annual growth and profit projections depend heavily on the performance of relative few accounts and people.

Throughout the country in companies that have been excellent examples of "customer focus," a monumental breakdown is occurring thanks to the *Body Snatchers*. The large block of quality time that should be spent with these important customers is on the edge of a cliff these days. We must be *reawakened* to the critical value of quality time spent with the priority customers that can make or break our successes.

In most organizations, the need to develop additional information or data about these key customers is nil. Yet, operations executives are finding more and more distractions and reasons to be studying data on their computers, and spending less and less care for the customers who count. The prism has shifted to an inside, egocentric viewpoint versus the external one it should be.

This is not unlike the trouble caused when Nero was engaging his fiddle whilst Rome went aflame.

No data mining from data warehouses is needed for key customers. These tools are used for people you don't know—not so much for the good customers, whom you should consciously know in every way, shape and form. But the truth is these good customers are becoming strangers in the process of data mining. They are getting lost in the shuffle as we pore over data about potential customers, glued to our screens and devices.

It has become too easy for customer care people to stay in their Pods to study data, make plans, exercise their computers and think about the bottom line instead of the top line. Contact with key customers has become an occasional e-mail or voicemail or chance meetings at some industry forum. The excuse is this treatment is necessary, because these "customers" are busy and don't want to be bothered." In addition, it's hard to get a hold of them or get them interested in our initiatives. What a hell of a way to handle your most important assets!

Never, ever take your key customers for granted. Love 'em. Keep in touch with 'em. Hold 'em dear.

My focus throughout my career was to find unique ways to spend hours with key customers, not minutes or seconds! I used the golf course, luncheons and dinners, our home seminars and retreats, and personal visits to the customers' premises in order to see them face to face. I know this is the old-fashioned way but I always thought it was important that customers knew that people in our company cared about them, from bottom to top.

I can't count how many times these customers would tell me "you are the first CEO I have ever met." I could always tell that they felt this gesture was very important to them, and it also made me feel good that I had initiated the contact to thank them for their business. See, e-mails and voicemails just don't get it done for me.

A close friend told me several times that there is no higher compliment than to invite customers, associates and other business contacts into your "home." I never forgot that and tried to create an atmosphere with our employees, customers and outside agents—even in the company facilities—of "entertaining at home" in all of our corporate social gatherings.

—Cary Blair

The world we live in today is hurried and we constantly hear from business people with whom we come contact that they just do not have the time to develop new customers, provide customer service to existing customers, attend meetings or even volunteer for charitable or non-profit causes. Certainly, in this time of frequent business consolidations, successful firms are rapidly acquiring other lines, organizations, and even their competitors. Resultantly for these and other reasons, sales people are asked to handle larger loads, in service and new sales. Time management is critical. They get the idea that if they grab onto every gizmo they can find, this will make them more efficient, when in fact, often, it does the reverse.

Remember, you simply have to <u>find time for the 10 percent that pay 60 percent of your salary or commissions or your bonus income!</u> Everyone who is successful in his or her life is busy, but you better be out there with current customers saying "thank you" and asking, "what else can I do for you?" You may even be shocked that your customers like the attention and want to recommend you to other business friends.

Time management is critical to your success . . . never take a customer for granted! Find ways to spend hours with key customers, not minutes or seconds.

"We thought we'd like to look around to see how the other half lives."

CHAPTER TWENTY-ONE

Shape up and ship out!!

IN THE MILITARY, when you have been confined to a desk job and are reassigned to combat duty, your mental and physical skills need a close look and some honing so you can be effective in all of your battle encounters. They call this "shaping up and shipping out."

The same is true for the Poddies who have been confined to their "Pod-jobs," deposited in their cubes behind screens playing with their gadgets. If you are entering or re-entering the world of "customer care" a skills inventory is necessary so you are in your best physical and mental shape to encounter an ever-changing and demanding customer.

Today, you do not have to be out of circulation for long to lose the "feel." Ironically, most of the tools at your disposal today are designed to take you out of circulation, not put you face-to-face with people, real customers.

Here's a skills checklist for customer care candidates:

- Meeting Other People . . . Introductions

- Looking Good . . . Dressing the Part

- Relating to People . . . Being Prepared

- Keys to Conversations

- Intelligence Gathering

- Communication Skills . . . Public and Private

- Gadget Etiquette and Other Do's and Don'ts

- Business Meetings

- Business Entertaining 101

- Promises . . . Delivery

- Personal Follow Up

- A Customer for Life!

This checklist is certainly not intended to be all-inclusive but does identify some of the most important qualities of a customer care specialist, whether in sales, marketing, branding, advertising or public relations. Failure to adopt and practice these skills spells trouble for neophytes and veterans alike.

Meeting Other People . . . Introductions

Customer Care means meeting new people continuously, day after day, week after week. You will be making many introductions and struggling to remember names and titles. There is no magic formula for remembering names. Good sales people use various methods. A popular one has always been to try to associate a name with something about the person's physical features or through visualizing the name written down boldly or in some other context.

When a person gives you his or her name, don't just get into a mind drift and think of the first thing you want to say. After the person gives you his or her name, repeat that name slowly . . . "Bill Austin . . . it is nice to meet you."

People have told me over the years that I was good with names. In fact, our President and I were very good with names. The only problem was that often it took both of us to remember the full name. He would come up with the first name and then I could get the last name or visa versa. That worked great when we were together but unfortunately, we rarely called on customers together or traveled together. It was necessary to find some other method. My method involved a "memory minder" that I normally always carried in my jacket pocket. After meeting new people, I would concentrate on the names and record them on my memory minder at the very first opportunity and these opportunities normally come up frequently. At every opportunity, I would pull out the memory minder and review the name, which many times I would spell phonetically for future introductions, if the name were difficult to pronounce.

—Cary Blair

Introduction of people is also an important skill required of customer care advocates. A good tip is to always try to help the people around you with introductions. Introduce yourself at the first opportunity, many times; this can be a great help to everyone around you. "**HELLO, I AM ZYDRUNAS ILGAUSKAS, EVERYONE JUST CALLS ME 'Z'.**"

The Second tip is to always keep a good reference manual at your fingertips to help with difficult introductions that you know are forthcoming. We suggested earlier *The Etiquette Advantage in Business*, by Peggy Post and Peter Post.

Looking Good . . . Dressing the Part

The business dress code at the beginning of the 21st century is in disarray. Poddies have had a lot of undue influence on the casual nature of current business "attire" in many companies. "Business Casual" is the new vogue, especially in those companies driven mainly by an "internal focus" and those companies with high Poddie populations.

The arguments for business casual range from clothing expense savings for the employee, higher productivity, and better morale. If the truth is known, most business casual policies were driven by techies and bean counters when their particular skills were in short supply and there was high demand in their sectors. More and more the question of dress code would come up when human resources was recruiting. After getting shut out a few times because of a lack of a business casual policy, H. R. put out the word that business casual was a "must do."

Business casual is an appropriate name for the look and feel of this approach to dressing. It creates a real dilemma for customer care employees who should have an "external focus" everyday. It also makes external visitors uncomfortable and uncertain about how they should appear on your premises. The current rule is to dress to the standards of the people you will see that day.

If the highest standard is business casual, that dress is appropriate for you. If they normally dress in business attire, a suit with tie is demanded of you. The reoccurring problem with business casual is surprises. You should be in jacket and tie, and you find yourself dressed casual. Oh well, you better also bone up on apologies.

I once had a client that was—with the emphasis on was—on the Fortune 100 list. It had a grand record of accomplishment and innovation and was known the worldwide. Then something happened.

The CEO decided that the firm should take "business casual Fridays," and make that de rigueur the other four days of the workweek. This would lighten things up and put everyone on an equal footing. But what kind of message was being sent here??

Through the 1990s this firm began to progressively fail. People just didn't seem to care that much; they were so relaxed and informal. A picnic, laissez faire philosophy prevailed about everything—not just the clothes. A lot of people just took a slob approach to what they wore.

Today the company is just a shadow of its former self, withering on the vine of commerce, trying to survive. The every day casual approach has been changed by another CEO to more business-like, professional attire. But the lazy boy and girl philosophy of the past is still imbued throughout the company.

Think twice about the notion of informal business dress. There could be a deeper connotation and that might not necessarily be good for the professional discipline it takes to keep an organization on its feet and successful.

We think there can be some happy medium here, with men wearing at least a sport jacket with a good pair of trousers, and women wearing a jacket, slacks or a skirt as part of their dress.

Of course, if you or one of your people need to present testimony before Congress, don't you think a suit and tie for men and similarly professional clothing for women are appropriate? If you are at a non-black tie but still somewhat formal business gathering, don't you think the same is true?

—Ron Watt

We are not fans of this new vogue. Business casual promotes a "indifferent" attitude and "feel" to your company and your job to you customers. It is hard to police in the workplace and creates a lax attitude. Business casual changes the "culture" of the company. Business casual fits right in with being trapped in a cubicle, behind screens, working gadgets, *making your company weak, complacent, tedious, predictable, boring, loutish, anti-social, changeless and risk-averse.*

If you are a customer care person, be prepared for your day with your customers and other interesting people. When visiting business casual companies, a handsome jacket, shirt and slacks will carry the day. Somehow that jacket always says, "I mean business!"

"I believe this concludes our crisis preparedness meeting."

CHAPTER TWENTY-TWO

Relating to People: Be Prepared

WE PREACH CONSTANTLY that high tech and the addictive gadgets have created three generations of anti-social people. Call them "closet executives" or "Poddies;" they are ill prepared for meaningful and constant contact with human beings, especially customers and other social animals that thrive on this contact. A large portion have been snatched, cloned and left in their Pods to wither away on "same-o, same-o."

If you are going to enter this "outside" world, remember that customer care specialists create the excitement in people's lives. Be prepared for that mission. Your mouth may be flapping most of the day and your head nodding. What comes out of your mouth better be good and grammatically correct and interesting and decisive.

Good social skills can be learned or re-learned but make a good effort if you have been out of the mainstream for awhile. Join and volunteer inside and outside the office, get back into your speech class, if you need it, or become a member of Toastmasters; and try to mix with a varied group of people from different occupations and social strata. It will help you become the "social animal" you dreamed of being. You can be a meaningful mentor, tutor and trusted advisor to prospects, customers, fellow employees and others with just a little thought and preparation.

Keys to Conversations

We reference what you probably suspect from our business etiquette bible, *The Etiquette Advantage in Business, Professional Skills and Professional Success* by Peggy Post and Peter Post (Harper Collins):

In even the most unconventional fields of business, is being a good conversationalist really so important? In a word, yes. How you speak, the quality of your voice, and your choice of words are fundamental to how you're perceived. People who have poor grammar, who are indifferent listeners, or talk mostly about themselves are seen in a less-than-positive light. Whether you're making a sales presentation or chatting with your supervisor, the ability to reach and influence a listener is one of the most valuable assets you have.

—Cary Blair

Being a good listener, of course, can never be underestimated. Like Calvin Coolidge once said: "No man ever listened himself out of a job [women included]."

Poor speaking, on the other hand, can have serious consequences: Eighty percent of executives questioned in a recent study by a Midwestern university cited a lack of communication skills—not technical expertise and overall performance—as the reason employees were held back in their careers. Compounding the problem is the boss who won't hesitate to tell you when your job performance falls short, yet is reluctant to criticize your way with words. That said, it's usually up to you to determine whether your speaking skills need work. Find out by broaching the subject with a co-worker you're close to; he, too, may be uncomfortable offering criticism, but when you assure him he'll be doing you a favor, he may very well consent."

We view conversation as a series of well thought out questions and comments and intensive listening to the responses.

Conversation Tip: Try to begin your questions with an interrogative (Who?, What?, When?, Where?, Why? and How?) Questions framed this way always solicit a narrative answer, not "yes" and "no."

"Mr. Jones, did you have a good year in the business?"

. . . "Yes."

"Mr. Jones, **how** did your business fare this year?"

. . ."We had an okay year but we are seeing more and more intrusion into our business by foreign competition, and our company has compressed margins. The price of raw materials continues to rise forcing us to put two price increases in this year placing us in a Catch-22 situation. Our future is not as bright as we would like."

If you are a good listener, you can learn a lot with one good question.

Gadget Etiquette and Other Do's and Don'ts

Just think about your last commercial airline flight and the cell phone conversations that you were privy to before you took off and immediately after landing. These must be life or death calls to disturb four or five rows of passengers.

What about your last meeting when participants pull out their BlackBerry and proceed to tap, tap, tap on the screen right in the middle of you presentation?!

Have you made an important phone call to a customer or vendor and hear the click, click, click of a keyboard in the background? Your call must not be as important to the receiver as you thought it was.

How many times have you been to a concert, an opera, a meeting, a restaurant, or even a funeral when the song of a cell phone interrupts your concentration and that of others?

Gadget etiquette is quite simple. Do not disturb other people around you with your gadgets in public places and always show respect and be a good listener at meetings where you are a participant. Be careful that you don't convert from a victim to a perpetrator. Find a quiet place for your cell phone conversations; the people around you do not want to hear these calls unless they are industrial spy operatives. Do not take phones where they do not belong and present your BlackBerry only when asked to check your calendar or make an entry. Don't become a gadget jerk, because it only reflects on who you think is the most important person in this world, YOU!!

What did you say, Pierre? Press 1 for filet mignon?

"Oh, Lord! Not another wine–and–cheese party!"

CHAPTER TWENTY-THREE

Entertaining Customers 101 (A Required Skill of Social Animals in Customer Care Disciplines)

RESTAURANT OWNERS WILL TELL YOU that business entertainment has waned in the last ten years. Some of the downturn simply follows the economy; this expense is also one of the easiest for management to quickly cut when company performance is also less than expected. Again, this is a characteristic of "internally focused" companies where customers are viewed more of as a "necessary evil" rather than a "valued asset." In these companies the first cuts are always in Public Relations, Advertising, Marketing, Sales and Branding.

There is absolutely no need to heap lavish gifts, dinners or trips on your key customers on a regular basis. However, customers do want to know you care, especially those that make large contributions to the top and bottom lines and those up-and comers that strive for the top spot.

A customer care program should be consistent and constant in bad times and good.

You must find ways to say thanks where you get in front of your customer, face-to-face. A quiet dinner before a sporting event or play, a day of fishing or golf, are old standbys, and they work. Some of these events offer an opportunity to include spouses, a much overlooked asset to your agenda.

Try a new outing style from time to time. Put together a small group of clients or prospects in a "retreat" where half of the day is recreation and the other half concentrates on business problems and opportunities and participants feel like they take valuable information back to their businesses. One of the easiest and most effective is a review by an outside expert of new laws or regulations that have been enacted that have broad implications to various businesses. Compliance and other Human Resource issues are also good topics to dig into at these retreats.

Find your own ways to keep entertaining expense under control, or you will eventually be "told" how to make the cuts. Scout out inexpensive restaurants and other venues that present a value equation and present an attractive new setting for a customer dinner.

Hopefully, some of your selections allow your customers and prospects to see these venues for the first time. But make sure that this is not your first time on the premises. Always try the restaurant alone first—to judge quality of the food and service. Yet be creative in alternative places that your customers might enjoy and not forget soon.

Breakfast and lunch entertainment is more popular today, again an expense consideration. These types of meetings generally result in an "all business" atmosphere and normally tight time restrictions. But your preference still might be a dinner meeting that is more casual and where time is less of a factor.

I always liked to entertain around golf and other sporting events. It was a nice way to get quality time with a customer to make commitments, solve problems and answer important questions that the customer had on his or her mind

As a side thought, it also is a more informal setting that allows for more casual dining and conversation. In this setting fewer of your shortcomings in business etiquette are exposed than at a formal dining situation. Save the more formal dining encounters until you have done some homework and practice.

—Cary Blair

The best way to be prepared for formal entertainment is to get ready with a group of your peers and staff, and have some fun doing it.

Assemble your colleagues and their spouses and do the ultimate dinner under the direction of a competent "entertaining or etiquette" consultant in your area. Have a fun five-or-six-course meal with instruction, including invitations, wine selection, food selection, utensil use and some of the common "do's" and "don'ts" in this formal setting. This educational exercise will be invaluable to you and your guests for the rest of their lives. There is nothing like doing it "live" to leave a lasting impression. Don't be afraid to make mistakes in this setting, as long as you learn from them.

Business Meetings

Many of my business acquaintances have commented over the years about my punctuality around meetings. I am normally always early to all of my appointments and some times it drives my wife crazy when I want to "get going" early for a casual dinner appointment or busines-related dinner. She often comments, "We will be the first ones there again." But I just hate to be late.

My training categorizes "late" as a total disregard for other people's time. Being late to appointments or meetings is a bad habit with some executives and managers, young and old.

People entering a meeting late have always been an irritation to me, even if I wasn't the presenter and I would always comment on this trait in performance review sessions if it happened frequently. If you expect to be dealing with the "outside" world, don't fall into this habit; always be early even if it means 10 or 15 minutes sitting in your car reviewing your agenda.

Occasionally, unexpected situations occur that are going to make you late to an appointment. Make sure you always phone ahead to the client or administrative assistant as soon as possible. Offer to reschedule if there is insufficient time to accomplish the meeting agenda.

Finally, try to be very considerate of your staff and others when scheduling inside meetings. Check travel schedules and other commitments so your meeting can be scheduled at their convenience rather than yours. Also, try to make most meetings early in the day rather than later.

Many of your young executives have young, active children who participate in sports and everything imaginable extracurricular. There are family demands on your execs' time, and most of those demands are early in the evening. I always tried to treat my staff as customers, focusing on their needs as opposed to mine. In my 12 years as CEO, there was rarely a time *when it was necessary to call an "urgent meeting" after normal business hours.*

—Cary Blair

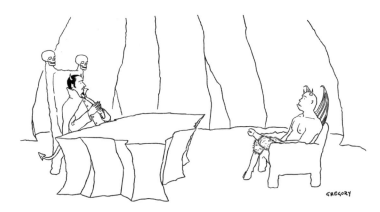

"I need someone well versed in the art of torture—do you know PowerPoint?"

CHAPTER TWENTY-FOUR

How not to screw up a meeting presentation.

PRESENTING AT A MEETING is a real dilemma in this world of technology. *What piece of equipment do I use and what software will give me the best illustrations? Do I need a backup take-away from the meeting?*

A good rule of thumb is to use equipment and software that you are very familiar with, because there is nothing worse than a PowerPoint presentation gone afoul. How many times have you been in a meeting when a call has to go out to an administrative assistant for "help"? This normally is the result of last-minute preparation, with little practice time. Don't press your luck in these important situations by trying things new that you are unfamiliar with. Just use a presentation technique that complements your style, sometimes a simple whiteboard with key points is all that is needed, along with a printed leave-behind piece. Keep it simple.

Meeting Tip: Keep meetings as short as possible, provide an advance agenda when possible and stay on time. Many meetings find their way "into the weeds" when schedules are not adhered to. Sometimes "talkers" take you to never, never land. Run your meetings efficiently and get to the point! Make sure that other presenters know their time allowance. This sets a favorable expectation for future meetings for participants who also have time to manage.

Please interrupt me at any time if you have a question.

Speaking Skills . . . Public and Private

You don't have to be an oratorical wonder to be an effective speaker. Plainly, you should just be yourself, with your unique personality. Have a high regard for the King's English but there is no reason to become rigid and pedantic in phrasing your thoughts to a small group or a large one.

Stay on message. Have a few key points you wish to make, and get off the stage sooner than later. Make sure there are some moments for questions and answers, where you have the opportunity to reach people who are most interested in what you have to say. If you don't know the answer, tell them you'll get back to them shortly after the meeting.

Speeches are not that unlike conversations. People communicating with one another. Not too dissimilar from a book author and his reader. Ultimately, it means one person talking to another, even though there maybe a few dozen or a few hundred people present.

Look around the room as you talk, not just to one person. Look at the people who are alert and seemingly in connection with what you are saying.

Practice what you are going to say. But never read your speech, unless it has to be extremely precise, such as testimony before legislature or some other form of oral communication that will be recorded for posterity. Mostly, people should be able to give a speech the way they talk naturally. Just don't ramble. Be able to walk away from the address with the idea that the vast majority of the people in the audience "got" what you had to say.

Like anything else, the more you speak formally, the better you'll get at it. Some people get self-conscious when they have to make a talk, but if they would "read" and work with their audience that would never be an issue. Instead, the speaker would be thinking more about the audience and what that group needs to hear and know rather than how he or she is coming across.

Just remember, some people are more naturally dulcet-voiced speakers than others. Some people can seemingly give a good talk, but actually say nothing. Insincerity, even from a good-on-his-feet speaker, is always worse than someone who is not so "smooth" but truly has something to say.

Intelligence Gathering

If you are going to practice "customer focus," intelligence gathering is an essential skill. This is an opportunity for you to use some of your "Pod Skills," because the Internet and your data warehouse can provide you with valuable information about your customer or prospect. The Internet often provides considerable information about the company and its mission, its competition, and vast details about its product and service offerings. If the company is public, a wealth of information on the company and its key officers is available, as is its financial performance. On the Internet, you'll see everything from the woes and marketplace pressures to the opportunities and accomplishments of the company.

Many factors are at play when you meet new prospects or even visit current customers. The global economy makes the marketplace dynamic and you have a need to know all you can get your hands on. Don't be lax in your preparation for new encounters. The tools to prepare are all around you . . . and don't forget the local library as well, with its abundance of periodicals, references and catalogues that pertain to various industries and companies.

In addition to the wealth of information on the Internet and in libraries, *The Wall Street Journal* is an essential tool, as are many other general daily, weekly and monthly business publications and journals. Have your assistant scan them frequently for articles that help you understand the ever-changing world of business around you.

The last thing a prospect or customer wants to hear in response to his question is "I didn't see that" or "sorry, I must have missed it" or "I haven't had time to catch up on my reading."

If you intend to become a "trusted advisor" to your clients and create customers for life, gather your intelligence and be well prepared for business encounters!

Promises and Delivery

It is a rare case that you will ever be in a situation where you can control the fate of a product or service from beginning to end of a sales or promotion cycle unless you own a one-person business.

Be realistic about your promises because you are expected to deliver "on time, every time" by your clients. Remember that realistic promises often out-pace the competition. Let the "wows" be a surprise, unexpected event that does not set precedence for every transaction. Strive for excellence and consistency, which many of your competitors might not be able to do.

There is an old saying in the South that "if you find a turtle sitting on a fence post, chances are very good that he didn't get there by himself!" It is well to remember that you will need help getting to lofty places in your career. Delivering on your promises in large part depends on your fellow employees. Treating them as customers is not only considerate but also a wise thought you won't regret.

Every employee likes to be recognized for a job well done. Share the "wows" with them from the customer. And, in turn, let the customer know who played key roles in a successful outcome. Apply the same customer care techniques on fellow employees that you would use for an external customer!

Mistakes and unfulfilled promises are the greatest "downers" in the customer care field. Sometimes these problems are not *your* fault. Customers sometimes become irate and unreasonable, like any frustrated people you encounter. Even so, most will listen and calm down if you make a reasonable, truthful explanation of why the ball was dropped. A quick and thoughtful apology is in order from you on behalf of everyone involved. Following that, the next order of business is getting the promise fulfilled as quickly as possible. Normally, customers are not lost when immediate attention is given to the problem.

The proactive approach to a mistake is to acknowledge it instantly, correct and learn from it.

It is one thing to make a mistake, and quite another thing not to admit it. People will forgive mistakes, because mistakes are usually of the mind, mistakes of judgement.

But people will not easily forgive the mistakes of the heart, the ill intention, the bad motives, the prideful justifying cover-up of the first mistake.

—from The Seven Habits of Highly Effective People by Stephen R. Covey (Simon & Schuster)

"And now at this point in the meeting I'd like to shift the blame away from me and onto someone else."

Personal Follow-Up

The authors of this book feel like they personally keep the producers of stationery goods in robust business shape. We admit to being incessant personal note writers. Our e-mails will go unanswered for days, but not the personal acknowledgement of compliments, congratulations, new business orders, gifts and the like. These notes are penned immediately and get to the point. This is simply "old fashioned customer care" and even the "Poddies" reluctantly agree that they like to receive a personal note when they have done something special, not some misspelled e-mail!

Our society is so imbedded in *dullness and drudgery* that some of the simple, effective follow-ups have gone by the wayside with the dawning of the new "closet executives." Each of us is constantly reminded how important the basics of business and business etiquette play in a company's success and for the people that represent it.

I will never forget a note that I received in 2000 written inside of a gift book on business etiquette, penned by one of the authors. After you have read the note, just think about the different impression and feelings that you would get if you received this by e-mail instead of as a personal hand-written note:

For Cary Blair,

Greetings to one of the most considerate executives I've had the pleasure of working with. You could have written this book!

Best Wishes,

Peggy Post
February, 2000
(co-author, with her brother-in-law Peter Post, of
The Etiquette Advantage in Business—Personal Skills
for Professional Success—Harper Collins)

Make your clients and customers, fellow employees and friends feel special with a personal note whenever you can. You will be pleasantry surprised at their reaction.

Customers for Life

Retention of customers and landing new ones is extremely difficult today with the increase in foreign competition, Internet sales, mail-order sales, sales promotions on television and gads of other impersonal sales pitches from every quarter. Face-to-face meetings with prospects and clients are harder and harder to come by these days. But don't cave in. Smart business executives still always find time to see their customers, and usually, the customers are happy to see them. Everybody wins.

In yesteryear as well as two centuries from now, they will always come out the most successful of their counterparts because they understand the simplest prospect of human interaction: Get to know your customer, get to be his or her friend. Don't hide behind devices that are means of implementation but not means to an end.

Social skills and tried and true sales and business methods have been lost in a world of screens and gadgets.

Learn or refresh these etiquette skills, and we know you will see an upturn in your success. Despite all of your efforts that emanate from the cubicle, customers, prospects, friends and neighbors can instantly determine when "you ain't got no class" and "you ain't going to be successful, and you just don't care."

It may be safer in a cubicle, Dr. Bennell.

CHAPTER TWENTY-FIVE

Horrifyingly, this is real life,
not science fiction.

WARNING: DR. MILES BENNELL (1956)

The world is being taken over by body snatchers, emerging from giant seedpods. People are being replicated in Pods and their real bodies have died. It's not much different than the walking dead—people without emotion, souls, hearts or love.

WARNING: CARY BLAIR AND RON WATT (2005)

The world is seeing the return of the body snatchers. Three generations of young people have been captured and cloned in cubicles and it's as if their real bodies and emotional drive have died. It's not much different from the walking dead, as they turn themselves and their companies into organizations of weak, complacent, tedious, predictable, boring, loutish, anti-social, changeless and risk averse hominids!

Our working lives and our personal lives should be fun and exciting, not encapsulated in dullness and drudgery. We should have many opportunities for personal successes in our life and also have many opportunities to create victories for our company. Our opportunities dwindle rapidly if we allow our lives to be possessed by hardware, software, cell phones and other electronic gadgets. This is especially true when this is time that should be spent on our personal development, with "customers" or with our families and friends.

Our idea of fun at work and fun at home should not include hour after hour in a cubicle or at a desk behind a computer screen. That is just too much like science fiction for our taste. What happened to hobbies, games,, biking, walking, thinking, reading, getting close to nature and so many other interests in which we can partake? Where has our creative spirit gone? We have become glued to our gadgets and our screens as pure and simple addiction. This is what shapes our lives as humans? This is how shallow we have become? This is how disinterested in the world around us we are?

This is real life not science fiction! Human beings are social animals. It is normal to enjoy the company of other human beings. It is normal to discourse with other people face-to-face. It is normal to try to help others with problems or opportunities, face-to-face. It is normal to ask customers for their business face-to-face. Using e-mail to say thank you just doesn't get it done!

We have seen several generations of young people turned into "closet executives" by their families, elementary and primary educators and universities, and even the companies they work for. Face-to-face contact has been taken away by the allure of computers and other flashy electronic machines and by the multitude of possessive software programs. We see more and more young "robots" addicted to this modern technology. Given enough rope, these gadget robots will eventually reshape "customer care."

Most of the companies that are heavily infested with these Poddies (People of Dullness and Drudgery Imbedded Everywhere) deserve their ultimate fate— becoming average, so-so, not bad, typical, mediocre, ordinary and run-of-the-mill. Nothing.

Let's build our personal lives and business lives around thoughtful time management . . . quality time with people and responsible and logical time for technology. And then let's enjoy our lives out of the Pod!

Human happiness and human satisfaction must ultimately come from within oneself. It is wrong to expect some final satisfaction to come from money or a computer.

—Dalai Lama, The Pathway to Tranquility (Penguin Compass)

"Stand aside, Gruenwald! It's the computer I'm blowing away!"

"He's dead. Would you like his voice mail?"

Where did all the Poddies go?!

The Authors

Cary Blair and Ron Watt

Cary Blair (C.B.) is the retired Chairman and CEO of Westfield Group, a highly rated Property/ Casualty Insurance and Financial Services provider headquartered in Westfield Center, Ohio. He spent his entire 43-year business career at Westfield, retiring in 2003, after 12 years as CEO.

His first job with the company was in marketing and marketing was his love for his entire career. He was always described as a "visionary" and a "change agent" in the company and the industry. He was an advocate of an "entertaining at home" feel to corporate hospitality and often rotated executives to unfamiliar disciplines to round them out for leadership positions in the company.

He has given back generously to community, industry and to national trade associations. He and his company have set an example for business ethics and leadership in the region and have received numerous awards for that leadership and volunteerism. He serves on the boards of three publicly held corporations and is an advocate for his alma mater, Butler University, and for his adopted Ohio institution, The University of Akron.

"C.B." loves to "work the crowd," says "thank you" to customers in a variety of ways personally and loves high-performing people and never minds stepping out of the spotlight for others that are more deserving.

* * *

Ron Watt is the author of many books, including two recent novels, *The Big Egg* and *Dateline: Ubi*, as well as the memoir, *A Love Story for Cleveland.* He was one of the authors of the highly acclaimed *The Art of Public Relations.*

He was founder and CEO of Watt, Roop & Co., one of the largest public relations and marketing firms in the Midwest, which later became Watt/Fleishman-Hillard, after the merger with Fleishman-Hillard, one of the world's largest PR firms. His clients have included American Express, Interbank, Citicorp, Marconi, Owens-Corning Fiberglas, the Cleveland Browns and Indians, KeyCorp, TRW, AT&T, many healthcare institutions and healthcare service and product companies, companies in the insurance and financial services fields, and trade and professional associations.

Watt is a Fellow of Public Relations Society of America, and former chairman of the Counselors Academy of PRSA, a body of executives and owners of more than 1200 PR firms worldwide. He is a member of the Cleveland Advertising Association's Hall of Fame and the Arthur Page Society, a group of leading senior PR practitioners in corporations, institutions and agencies throughout the U.S.

Besides his writing endeavors, Watt continues in the public relations, marketing and advertising fields as CEO of the Ronald Watt Companies and Watt Consulting LLC.